An Assessment of Mercury in the Environment

A report prepared by the
Panel on Mercury
of the Coordinating Committee for Scientific and
Technical Assessments of Environmental Pollutants

Environmental Studies Board
Commission on Natural Resources
National Research Council

NATIONAL ACADEMY OF SCIENCES
Washington, D.C.   1978

NOTICE

The project that is the subject of this report was approved by the Governing Board of the National Research Council whose members are drawn from the Councils of the National Academy of Sciences, the National Academy of Engineering, and the Institute of Medicine. The members of the Committee responsible for the report were chosen for their special competences and with regard for appropriate balance.

This report has been reviewed by a group other than the authors according to procedures approved by a Report Review Committee consisting of members of the National Academy of Sciences, the National Academy of Engineering, and the Institute of Medicine.

This study was supported by the Office of Health and Ecological Effects, Office of Research and Development, U.S. Environmental Protection Agency, Contract No. 68-01-3253.

Library of Congress Card Catalog Number 78-51128
International Standard Book Number 0-309-02736-5

Available from
Printing and Publishing Office
National Academy of Sciences
2101 Constitution Avenue
Washington, D.C.   20418

Printed in the United States of America

082660

PANEL ON MERCURY

Frank M. D'Itri (Chairman), Michigan State University
Anders W. Andren, University of Wisconsin
Richard A. Doherty, University of Rochester
John M. Wood, University of Minnesota

Staff Officer: Adele L. King

COORDINATING COMMITTEE FOR
SCIENTIFIC AND TECHNICAL ASSESSMENTS
OF ENVIRONMENTAL POLLUTANTS

Ian C.T. Nisbet (Chairman), Massachusetts Audubon
  Society
Ralph C. d'Arge, University of Wyoming
John W. Berg, University of Iowa
Russell F. Christman, University of North Carolina
Cyril L. Comar, Electric Power Research Institute
Eville Gorham, University of Minnesota
Robert C. Harriss, Florida State University
Delbert D. Hemphill, University of Missouri
Margaret Hitchcock, Yale University
Robert J. Moolenaar, Dow Chemical Company
Jean M. Morris, E.I. duPont de Nemours & Company
Peter N. Magee, Fels Research Institute

Ex Officio Members (Panel Chairmen):

Julian B. Andelman, Panel on Low Molecular Weight
  Halogenated Hydrocarbons, University of Pittsburgh
Patrick L. Brezonik, Panel on Nitrates, University of
  Florida
Frank M. D'Itri, Panel on Mercury, Michigan State
  University
Robert J. Livingston, Panel on Kepone/Mirex, Florida
  State University

Staff:

Edward Groth III, Environmental Studies Board
Adele L. King, Environmental Studies Board
Charles R. Malone, Environmental Studies Board, (June
  1975 - December 1976)

CONTENTS*

*The different units used for expressing concentrations of mercury in various media are so diverse that confusion sometimes results. The Panel on Mercury has chosen to use the official metric system of units uniformly throughout this report. To facilitate comparisons, however, a table of units is presented in Appendix A.

THE STUDY ON SCIENTIFIC AND TECHNICAL ASSESSMENTS OF ENVIRONMENTAL POLLUTANTS

In early 1975 the U.S. Environmental Protection Agency (EPA) approached the National Research Council of the National Academy of Sciences to request a series of comprehensive scientific and technical assessment documents on selected multimedia environmental pollutants. These documents would be used by EPA as a basis for preparing Scientific and Technical Assessment Reports (STARs), which in turn would be used as the scientific and technical basis for possible regulatory action on these pollutants. EPA was anxious that these background documents use an environmental "mass balance" approach, i.e., that they attempt to account for the sources, sinks, and receptors for the pollutant as it moves through the environment. The initial two reports (Nitrates and Nonfluorinated Halomethanes) were also to explore the technology for controlling the pollutants and the costs and benefits of available means of control, whereas the subsequent two (Mercury and Kepone/Mirex/Hexachlorocyclopentadiene) were not.

Within the NRC, responsibility for the study was assigned to the Environmental Studies Board of the Commission on Natural Resources. When it agreed to undertake the study, the Environmental Studies Board identified two distinct objectives. The first was to conduct a series of assessments of specific pollutants, as required by EPA. The second was to draw upon the experience gained from conducting a limited number of such assessments, in order to address the broader methodological problem of how such assessments should be done and how best to use the limited resources of scientific expertise for environmental issues to meet EPA's expanding need for independent, critical scientific evaluations of pollutants.

A Coordinating Committee for Scientific and Technical Assessments of Environmental Pollutants (STAEP) was appointed to oversee panels, which would conduct the

assessments of specific pollutants, and to make
recommendations regarding the best methodology for
producing such assessments. The pollutants to be
studied were mutually agreed upon by the EPA and the
NRC; pollutants were chosen that posed particularly
complicated or difficult problems of assessment, given
current scientific and technical knowledge.

In September 1976, to help meet the requirements of a
consent agreement for documents on health and ecological
effects, EPA requested additional assessment reports
under the STAEP study program. The NRC agreed to
conduct two additional studies and, by mutual agreement,
mercury was chosen as the topic for one. The additional
studies differed from the initial two in that (1) the
performance period was shortened and the scope
consequently narrowed, and (2) the reports were to be
based primarily on a literature survey and review
prepared by another contractor for EPA's Review of
Environmental Effects of Pollutants (REEP) program.
Undertaking these additional studies would benefit EPA
by providing the best available expert judgment, and the
STAEP Coordinating Committee would gain the experience
of another method by which assessments could be
performed and would have a broader base upon which to
report on the methodology of assessing pollutants.

THE PANEL ON MERCURY

STAEP's Panel on Mercury was appointed in November
1976. The panel was charged to provide EPA with a
critical assessment of the available scientific and
technical knowledge on the human health and ecological
effects of mercury as an environmental pollutant. In
addition to assessing the ecological and environmental
health hazards of mercury, the study was to identify
areas in which the scientific evidence is uncertain or
too inconclusive to arrive at an assessment, and to
suggest areas where additional research is needed to
provide a sounder basis for future regulatory action by
EPA.

The panel conducted its study through a series of five
1-day working sessions from December 1976 through June
1977, including a public information-gathering session
where scientific and technical information from the
interested public was solicited and presented to the
panel. At its first meeting the panel determined that
the literature review (REEP-Mercury) provided by EPA was
not satisfactory as a primary reference for their work.
This unexpectedly expanded the panel's work to include
an extensive search and review of the general scientific
literature by the panel members and NRC staff. In the

past 6 to 7 years a significant volume of new literature, data, and analyses of mercury in the environment and episodes of human environmental exposures to mercury have appeared.

At their meetings, panel members presented the current work in their respective fields and pooled their expertise to discuss the interpretation and implications of this information. This report represents the panel's best assessment of new information on the health and ecological effects of mercury in the environment and how it bears upon the existing information. It also represents the panel's best judgment on related controversial scientific and technical issues. At the same time, the panel attempted to avoid being an encyclopedia of well-established information on mercury.

The panel appreciates the contributions and information of a wide range of people from universities, government, and industry. We especially appreciate the efforts of Dr. V.B. Vouk of the World Health Organization (WHO) in making available the galley proofs of the WHO's publication, Environmental Health Criteria 1, Mercury, for the panel's use prior to publication. I would like to note the dedication, prompt and conscientious attention, and close teamwork of the panel members who prepared this report, especially in the face of a very short deadline and an unanticipated expansion in their workload.

I would also like to thank the professional and support staff of the National Academy of Sciences-National Research Council, especially the panel's Staff Officer, Adele L. King, who made an exceptional contribution to the preparation of this document, Connie Reges for her secretarial support, David Savage and the entire Manuscript Processing Unit, and Philippa Shepherd and Judy Cummings of the Commission on Natural Resources Editorial Office.

Frank M. D'Itri, Chairman
Panel on Mercury

FINDINGS AND OVERVIEW

This study responds to a request by EPA for a critical assessment of the most up-to-date scientific and technical knowledge on the effects of mercury as an environmental pollutant. To base this assessment on the most complete, recent information available, each member of the National Research Council's Panel on Mercury updated one area of research to include the most recent publications in two basic categories: the impact of mercury on the environment and on human beings. The principal findings arising from this study are summarized below, followed by an overview that introduces the main themes to be found in the body of the report. Documentation for the findings can be found in the body of the report as noted parenthetically after each finding.

FINDINGS

General

1. Mercury compounds have no known normal metabolic function, and their presence in the cells of living organisms, including human beings, represents contamination from natural and anthropogenic sources. In view of the toxicity of mercury and the inability of researchers to specify the threshold levels of toxic effects on the basis of present knowledge, all such contamination must be regarded as undesirable and potentially hazardous. (See Chapter 6.)

Global Cycles

2. Although these estimates are tentative, the "pre-man" and present-day global mercury cycle models presented in this report (see Chapter 1, section entitled The Natural Mercury Cycle) support the following conclusions:

● The atmosphere plays an important role in the mobilization of mercury. From 25 to 30 percent of the total atmospheric mercury burden is anthropogenic. The residence time for mercury in the atmosphere is estimated to be 11 days.

● The mercury burden of lakes and rivers has increased 2 to 4 times pre-man levels.

● Total increases in oceanic mercury concentrations have been negligible. The residence time for oceanic mercury is 3200 yr.

● The mercury burden of freshwater and estuarine sediments has increased 2 to 5 times pre-man levels. The residence time for mercury in oceanic sediments is 2.5×10^8 yr.

● The mercury content of soil appears to have increased by about 0.02 percent. The residence time for mercury in soils is 1000 yr.

3. Flux rates between continental, oceanic shelf, and open ocean air masses and between shelf and open ocean water masses cannot be evaluated with current information. Models do not resolve the question of whether increases in mercury levels in watercourses, air, and soil have also elevated mercury levels in freshwater and land biota. The biological cycle of mercury is delicately balanced, however, and small perturbations in input rates and the chemical form of mercury can result in increased methylation rates in sensitive systems. (See Chapter 1, sections entitled The Natural Mercury Cycle and The Global Cycle of Methylmercury.)

Transformation and Uptake in the Environment

4. Identification of "hot spots" of anthropogenic mercury emissions to air, water, and land is valuable for establishing monitoring and health evaluation programs; it is not yet clear how valuable this identification is for projecting the mercury burden of aquatic and terrestrial biota and of the human population. Since the behavior of mercury is a function of its chemical form, data on emissions can be evaluated together with prevailing physical, geological, and chemical regimes for each region to identify sensitive areas. The release of mercury to the environment from coal-fired power plants, mining, and smelting operations is of special concern since these sources are currently uncontrolled and the use of coal is expected to increase significantly. (See Chapter 1, section entitled Man-Made Mercury Emissions in the United States.)

5. Mercury that is not recycled by industrial and other users is released into the environmental cycle and

becomes available for potential methylation. If the
element is not recycled, conversion to its original
geochemical form (for example, mercury sulfide) and
burial would be the best safeguard against
transformation into methylmercury. (See Chapter 2,
section entitled Mercury in the Lithosphere.)

6. Once the sources of anthropogenic mercury to
aquatic ecosystems are removed, the mercury content of
the sediments and aquatic organisms appears to decrease
slowly. The rate is a function of (a) the degree of
contamination, (b) the chemical form, (c) the physical-
chemical conditions of the system, (d) the hydraulic
turnover time, and (e) the rate at which the mercury is
either removed (bedload transport) or buried in the
sediments. (See Chapter 2, section entitled Mercury in
Natural Waters.)

7. Measurements of mercury levels in water and
sediments, though useful, are not sufficient to
ascertain the rates of methylation or uptake in biota.
Methylation rates in ecosystems are a function of the
mercury burden, bacterial population, nutrient loadings,
pH and redox condition, suspended sediment load,
sedimentation rates, and other physiochemical
conditions. Variations in methylation rates as a
function of each of these parameters in situ remain
unknown. However, cutting back on the point discharges
of soluble inorganic mercury compounds can significantly
affect methylmercury synthesis in polluted areas. (See
Chapter 1, sections entitled The Global Cycle of
Methylmercury and Man-Made Mercury Emissions in the
United States; and Chapter 3, Summary and Conclusions.)

8. For chemically sensitive waterways such as poorly
buffered lakes, the combined effects of acid
precipitation and increased emissions of mercury to the
atmosphere with subsequent deposition may pose a serious
threat if the correct biomethylation conditions are met
because more methylmercury would be available to the
biota. (See Chapter 1, section entitled Man-Made
Mercury Emissions in the United States.)

9. There is evidence that the bioaccumulation of
methylmercury into the tissues of higher organisms
(e.g., fish) is probably diffusion controlled. The
rates for bioaccumulation are so rapid (20 × 10^{-9} s for
methylmercury chloride to diffuse through cell membranes
into cells) that even low concentrations of
methylmercury in water lead to elevated concentrations
in fish. (See Chapter 3, sections entitled Kinetics of
Methylmercury Biosynthesis and Methylmercury Diffusion
into Cells and Summary and Conclusions.)

10. In highly contaminated aquatic systems fish and
other aquatic organisms may be at risk because they
efficiently bioconcentrate mercury; fish-eating birds

and mammals may also be at risk. In areas without high
natural or anthropogenic pollution, most animals do not
accumulate high levels of mercury and therefore do not
appear to be immediately at risk. However, the
sublethal effects on wildlife of long-term exposure to
low levels of methylmercury are currently undefined.
(See Chapter 4.)

Measurement and Monitoring

11. Analytical methods for measuring total mercury
in most biospheric and lithospheric samples appear to be
adequate. Measurement of total mercury in the
atmosphere and in the hydrosphere poses problems,
however, because the different mercury species measured
are currently defined by the collection technique and
the concentrations encountered are low. Nonetheless,
experienced analysts can usually determine total mercury
and methylmercury adequately. Strict quality control
and frequent interlaboratory comparisons must be an
integral part of any mercury analysis program. (See
Chapter 2 and Appendix A.)
12. It is difficult to determine the extent to which
an aquatic system is polluted by mercury because the
processes of methylation and uptake are so complex. At
present, the most effective indicator of both the degree
of mercury pollution and the potential hazards to humans
and wildlife is the mercury content of fish. In using
this indicator, factors that bear on the fish mercury
content such as age, species, and nutritional habits
must be taken into account. (See Chapter 2.)

Human Exposure and Risk: Establishing Guidelines

13. The most direct threat to human health from
mercury compounds is from consumption of methylmercury.
Therefore, the tolerance, guideline, or acceptable level
of mercury in food established for the protection of
human health should be set on the basis of the
methylmercury content of the food rather than the total
mercury content. (See Chapter 5.)
14. By themselves, mercury compounds in the
concentrations and forms usually found in the ambient
atmosphere and in drinking water do not contribute
significantly to mercury intoxication in human beings.
The levels of methylmercury in plants are also generally
extremely low, and do not contribute significantly to
the methylmercury burden in human beings, with the
exception of plants grown on contaminated soil or from
mercury-treated seed stock. (See Chapter 5, sections

entitled Air and Drinking Water and General Food
Surveys.)

15.   While all foods contain minute quantities of
mercury, only fish and seafood, and to a lesser extent
meat, present a potentially serious health hazard
because of their relatively high levels of
methylmercury. (See Chapter 5, section entitled General
Food Surveys.)

16.   The guideline for acceptable levels of mercury
in fish and other seafood is based, at least in part, on
the estimated average rates of daily fish consumption.
Although this protects the average consumer, it does not
take into account the subpopulations that consume large
amounts of fish and seafood or sport fishes from highly
contaminated waterways. (See Chapter 5, sections
entitled Mercury Levels in Fish and Human Uptake of
Mercury in Fish.)

17.   Preliminary data from animal studies suggest
that selenium has a protective effect against
methylmercury poisoning. However, the biochemical
mechanism for selenium-methylmercury antagonism is not
presently understood. Some information exists on the
selenium-mercury ratio in ocean fish, but similar data
supporting this ratio are scarce for freshwater fish.
Bishop and Boomer (1974) have presented data that
indicate that for freshwater fish there is no selenium-
mercury ratio. Currently, no data from exposed
populations document the protective effects of selenium
in humans. Therefore, present knowledge does not
justify modification of currently accepted guidelines
for total mercury or methylmercury in food on the basis
of selenium content. Further studies are needed to
establish the biochemical role of selenium and other
compounds that may interact with mercury and increase or
decrease its toxicity to living organisms. (See Chapter
5, section entitled The Effects of Selenium on
Methylmercury Toxicity.)

Effects on Human Health
(For a quantitative assessment see Chapter 6.)

18.   In human populations exposed to toxic doses of
methylmercury, signs and symptoms are dominated by
neurological disturbances. Currently, only relatively
insensitive clinical methods are available to evaluate
the effects of chronic low-level exposure to
methylmercury. Thus, current limits for detecting the
effects of methylmercury in human populations should not
be equated with threshold levels, for other more subtle
effects such as behavioral or intellectual deficits may
not be detectable by the clinical procedures that have

been used. More sensitive objective techniques for assessing neurotoxic effects of methylmercury need to be developed to supplement standard clinical neurological procedures. (See Chapter 6, section entitled Toxic Effects of Methylmercury in Adult Populations.)

19. Behavioral effects on humans of chronic low-level exposure to methylmercury are not fully understood, and information about possible genetic, reproductive, teratogenic, and carcinogenic effects of mercury compounds is incomplete and contradictory. Methylmercury has been reported to be weakly mutagenic in Drosophila, to reduce fertility in some rodent species, and to cause chromosome breakage in exposed human beings. Better data should be developed with which to evaluate the potential health significance of such effects. (See Chapter 6.)

20. Human subpopulations may be at increased risk because of consumption of very large amounts of fish or other seafood, or sport fishes from highly contaminated waterways; because of predisposing genetic or environmental factors; or because of differences in susceptibility at different developmental stages of the life-cycle. These subpopulations should be examined carefully. (See Chapter 6.)

OVERVIEW

The data so far collected on mercury contamination have affirmed that anthropogenic sources are raising the levels in air, soil, freshwater lakes and streams, and ocean estuaries if not in the oceans themselves. These increases are being methylated and translocated through the food chain. While the small amounts of mercury in most foods and larger concentrations from industrial exposures affect a few people, the greatest hazard from environmental exposures is to hypersensitive individuals who consume excessive quantities of fish or other seafood contaminated with methylmercury. The symptoms of acute poisoning have been documented in previous epidemics, and the precision and accuracy of analytical techniques for measuring organic and inorganic mercury in biological samples continue to improve. However, the data are still incomplete and the medical and technical proficiency are too limited to assess conclusively the effect of chronic exposure to low levels of mercury pollution on human beings as well as lower orders of the food chain and the environment. The mercury problem, therefore, clearly needs to be reassessed periodically to take advantage of new data and of the continuing refinement of chemical and analytical techniques. The approach of the present reassessment has been to review

the global occurrence of mercury and the mechanisms by
which it is transported and accumulated (Chapters 1-3)
and then to discuss effects--first on the ecology
(Chapter 4) and finally on human health (Chapters 5 and
6). The authors have attempted to be consistent in
reporting concentrations for biological samples on a
wet-weight basis throughout the report, unless otherwise
noted.

Occurrence and Bioaccumulation of Mercury

Mercury Cycles

The clearest conclusion to emerge from the panel's
review of the literature on the global cycle of mercury
is that substantial evidence remains to be collected
before definitive appraisals are possible. At this
time, the transport of mercury in the global cycle and
its sinks in the environment have primarily been
described by means of mathematical models based on
assumptions that have become questionable in the light
of recent data. For example, calculations of man's
influence on the global mercury cycle, which were based
on earlier measurements of mercury concentrations in
Greenland ice cores over time, need to be reevaluated.
Based on revised assumptions, a comprehensive synthesis
of old and new information on the global mercury cycle
in a "pre-man" and a present-day framework is
presented in Chapter 1.

New data introduced in Chapters 1 and 2 indicate that
the total atmospheric mercury burden is probably less
than previously estimated, and that approximately 25 to
30 percent is the result of man-made emissions. In the
United States the annual consumer commercial and
industrial mercury consumption for 1973 is estimated to
be 1.9×10^9 g. The total environmental losses
resulting from these activities, as well as from mining
and fossil fuel burning, is estimated at 1.5×10^9 g.
The range varies widely from one state to another,
depending on the population density.

Estimates discussed in Chapter 2 of the annual natural
flux of mercury to the atmosphere range from 25,000 to
30,000 metric tons (1 ton = 10^6 g), with elemental
mercury vapor the primary form that cycles from the
earth's surface to the atmosphere. Mercury's residence
time in the atmosphere is shorter than for any global
compartment, and is estimated at approximately 11 days
in this study. Elemental mercury is removed from the
atmosphere by rain and snow; and dry removal may also
occur. In the lithosphere, much of the mercury in

mineralized areas is bound as its highly insoluble
sulfide, cinnabar (ore).

Much has been learned over the past two decades about
the many kinetic factors that can change the speciation
of mercury in the aquatic environment. Calculations
show that ionic divalent mercury is likely to be formed
in well-oxygenated waters. This mercury species can
undergo several important reactions: (1) formation of
the highly insoluble mercury sulfide under anaerobic
conditions, (2) reduction to metallic mercury vapor
(degassing) under appropriate conditions, and (3)
conversion to alkylmercury (methylmercury and
dimethylmercury) compounds. Formation of
monomethylmercury occurs in sediments under aerobic and
anaerobic conditions and is greatly favored by low pH
(optimum pH 6.0). This is a key reaction since it
greatly increases the ability of mercury to cross
biological membranes, and methylmercury is rapidly
removed from the aquatic environment through
bioconcentration. Although many of the details of the
described cycle are as yet conjectural, the key elements
discussed are supported by current knowledge of mercury
speciation.

We do not yet have enough information to evaluate flux
rates for continental, oceanic shelf, and open ocean air
masses, or between shelf and open ocean water masses;
nor has the flux of alkylated mercury in natural
conditions been determined although order-of-magnitude
values are estimated in Chapter 1. We do, however, know
that mercury levels in watercourses, air, and soil are
increasing. Whether these increases have also elevated
the levels in biological specimens cannot reliably be
determined using models, but thermodynamic calculations
predict that under favorable environmental conditions
biological and chemical methylation of mercury will
occur, and the high levels of methylmercury found in
aquatic biota indicate that even extremely low
concentrations bioaccumulate rapidly.

Bioaccumulation Processes

Chapter 3 assesses the most recent research on the
chemical and biochemical mechanisms for methylation and
demethylation. Biomethylation is facilitated by three
methylating coenzymes in biological systems, but only
methyl-B_{12} is capable of methylating soluble inorganic
mercury salts to methylmercury and dimethylmercury under
both aerobic and anaerobic conditions. The rate of
methylmercury synthesis is determined by the available
concentrations of soluble mercuric ion species and
methyl-B_{12} compounds as well as by the nature of the

microbial community. The presence of demethylating organisms will allow a steady-state concentration of methylmercury to build up in an ecosystem, but the concentration would be lower than if they were less abundant. However, if, as the recent evidence indicates, the bioaccumulation of methylmercury into the tissues of higher organisms such as fish is diffusion controlled and very rapid, then even low concentrations in water can lead to elevated concentrations in fish.

Effects of Mercury on the Environment

Evidence presented in Chapter 4 supports the contention that, despite the often simultaneous processes of methylation and demethylation, fish and shellfish concentrate high levels of primarily methylmercury (usually 90 percent or more in that form) from the small quantities in the waterways. In contrast, non-fish-eating animals and birds usually concentrate less than 0.02 μg/g. In most unpolluted fresh waters, the top predatory fish such as bass, pike, and walleye may have natural levels of methylmercury ranging from 0.4 to 1.0 μg/g (wet weight) although concentrations as high as 2.0 μg/g may occur, whereas in highly contaminated fresh waters, they may average 10 μg/g, with some fish exceeding 24 μg/g. In general the degree to which the fresh water is contaminated is the principal determinant of mercury concentrations in fish. Nonetheless, age, weight, species, metabolic rate, and region of habitation also signficantly affect mercury levels.

Marine fish also concentrate primarily methylmercury. However, unlike freshwater fish, the levels are usually below 0.01 μg/g except for large carnivorous fish such as tuna and swordfish. They may have mercury levels that range from 0.2 to 1.5 μg/g. The quantity of mercury absorbed may be influenced by the position the species occupies in the food chain, the amount of salt in the seawater, as well as by the other factors described for freshwater fish. In marine fish the presence of selenium in levels equal to or exceeding the mercury content of the fish may reduce the toxic effects of methylmercury. For freshwater fish, the selenium levels may be too low to achieve this effect. High concentrations of mercury and selenium have been found together in the livers and brains of apparently healthy sea mammals; and the preserved umbilical cords of people at Minamata who were not affected by mercury poisoning had a 1:1 ratio of mercury and selenium. The mechanism of the protective action is not clear. The presence of selenium appears to immobilize the methylmercury compound but does not appear to speed elimination of mercury from the body. More studies of selenium should be conducted, and other compounds are also being investigated for mutual antagonism toward methylmercury.

Whether the levels of mercury in ocean fish are rising because of contamination cannot be determined because the historical data are insufficient to make accurate comparisons. However, evidence does show that fish from contaminated fresh waters may concentrate enough mercury to place local human populations at risk.

As the element concentrates up the food chain, it may affect the growth, reproduction, and behavior of organisms. The few data available indicate that most plants and other forms of life contain naturally occurring traces of mercury. Mercury poses a potential threat to the bottom of the food chain when it retards the growth of algae and zooplankton. Further up the food chain, insects and other invertebrates have a wide range of tolerances for various mercury compounds and concentrations.

Birds that prey on contaminated aquatic organisms such as fish also have high levels of methylmercury. Their food preferences and habitats, including migratory patterns, are the most important determinants of mercury concentrations. Life spans also have an effect: those that live longer have more time to concentrate mercury. In some birds the reproductive rate is affected.

Among larger vertebrates, premature births have increased among California sea lions since 1968, perhaps because of high DDT and PCB levels as well as mercury. Seals also have shown high mercury levels in other contaminated areas of the world.

Impact on Human Health

The information in the earlier chapters on occurrence of mercury in the environment and its effects on ecosystems forms the basis for the panel's assessment, in Chapters 5 and 6, of the routes by which environmental mercury reaches human beings and of the risks it poses to human health. These chapters also review the likely impact at the various stages of the life cycle, as well as possible genetic, reproductive, teratogenic, and carcinogenic effects.

Determinants of Risk

During the past few years substantial data have been gathered about mercury levels in human food, dietary patterns, and human mercury poisoning. The low levels of mercury in drinking water pose no threat to health. Nor do the minute quantities naturally accumulated in all foods. However, where fish have concentrated excessive amounts of methylmercury, sensitive individuals who consume large quantities may be at risk of poisoning. The groups at risk are indicated by fish consumption patterns, level of exposure to contaminated fish, and individual sensitivity.

A major risk is posed for fishermen and especially native Canadian guides in northwestern Ontario, where fish with up to 24 μg/g (wet weight) total mercury have been taken from the heavily contaminated Wabigoon-English-Winnipeg river system. A potential threat is also being investigated by the New Jersey Department of Environmental Protection in Hackensack Meadows.

Fish consumption varies with geographical location, race, and ethnic origin; and susceptibility to methylmercury poisoning appears to vary not only by individual sensitivity but also through the influence of factors such as nutritional status and concurrent exposure to other toxicants and infectious agents. These variables have not been evaluated yet in terms of possible additive or potentiating effects.

Toxicity to Adults and Dose-Response Relationships

To establish acceptable human and environmental tolerance levels for mercury, the toxicity to the central nervous system from chronic, low-level exposures, as well as other potential adverse genetic, reproductive, teratogenic, and carcinogenic effects, must be assessed. Results from study of human exposure to methylmercury compounds in Japan and Iraq have contributed significantly to the needed data base. A Swedish Expert Group made a detailed risk-evaluation of long-term human exposure to methylmercury compounds in the heavily exposed Japanese populations. The lowest observed blood mercury concentrations associated with the onset of symptoms was 340 ng/ml. Subsequently, with data from the large epidemic in Iraq, it was possible to determine the body burdens of methylmercury at the time of onset of various toxic symptoms and signs. The most sensitive index of toxicity was determined to be paraesthesia, which appeared at a mean body burden of between 25 mg and 40 mg mercury (as methylmercury in a 50-kg individual) depending on the estimated conversion factor. From these and other studies it appears that the critical organ system for adult human beings is the central nervous system.

While much has been learned from such studies, difficulties in quantifying clinical effects have prevented the determination of precise dose-response relationships at chronic, low-level exposure. Moreover, other more subtle effects may not be observed with current clinical techniques, and the minimal adult body burden at which methylmercury begins to cause damage to the nervous system must be assumed to be lower than can be demonstrated with the current limits of detection.

Toxic Effects on Developing Fetuses and Other Health Effects

Methylmercury poisoning has a proven toxic effect on developing fetuses; the fetal brain appears to be the most sensitive organ. In Japan prenatally exposed infants were often severely damaged, while their mothers frequently showed few or no symptoms. In Iraq it was possible to document prenatal transplacental exposure by measuring the maternal blood and hair mercury levels. There appeared to be clinically detectable fetal brain damage when the peak maternal hair mercury concentration rose above 100,000 ng/g during pregnancy (approximately equivalent to a blood mercury concentration of 400 ng/ml). Further evaluation of larger numbers of prenatally exposed individuals is likely to reveal that observable effects result from even lower maternal levels. In the exposed Iraqi population 31 percent of women who had apparent symptoms or signs attributable to methylmercury poisoning had maximum hair mercury levels less than 100,000 ng/g. In these individuals it has been suggested that there is a category of methylmercury poisoning in which symptoms such as paraesthesia, headaches, persistant pain, and weakness of the limbs predominate, with little or no evidence of neurological damage on clinical examination, and that these effects may occur with blood mercury levels well below 400 ng/ml.

Methylmercury has been shown to cause congenital malformations in mice and chromosomal abnormalities in rapidly growing plant root cells and lymphocytes cultured in vitro from methylmercury exposed humans. The significance of these observations for human health requires further evaluation. Overall, data on possible genetic, reproductive, carcinogenic, and teratogenic effects of mercury compounds are meager and somewhat contradictory.

Tolerances and Guidelines

Many important questions remain to be answered about
the effects of chronic, low-level human exposure to
mercury, particularly methylmercury. However, because
of the element's toxicity, temporary guidelines have
been established to limit the potential hazard,
particularly from consumption of mercury-contaminated
fish. In the United States the 0.5 $\mu g/g$ FDA interim
guideline usually only affects such large commercial
species as tuna, halibut, and swordfish. Some states
have imposed restrictions on the quantities of sport
fishes taken from highly polluted waterways (see
Appendix C).

Many other countries have also set recommended
guidelines to limit the consumption of fish contaminated
with methylmercury. These guidelines usually recommend
that pregnant women eat none because of the greater
danger to the fetus. In 1972, the World Health
Organization set a provisional tolerable weekly intake
of 0.3 mg total mercury of which no more than 0.2 mg
should be methylmercury. No firm basis has yet been
established for determining a safe standard for mercury
in foods. (See Appendix B: A Brief Review of FAO/WHO
Deliberations on Setting Acceptable Tolerances for
Mercury Residues in Foods, 1963-1976.)

Analytical Methods

One reason why only tentative or provisional
guidelines have been established is that more refined
analytical techniques are needed to measure selectively
the minute quantities of various inorganic and organic
mercury compounds in biological samples as well as in
water, soil, and air. Earlier analyses generally
measured only total mercury. However, the greater
sensitivity, selectivity, and reliability of current
techniques permit both inorganic and organic compounds
to be measured. In the past, analyses tended to err on
the low rather than on the high side because of lack of
instrumental sensitivity, accidental volatilization of
mercury, and failure to detect organic mercury
compounds. Current analytical methods are more accurate.
Errors are introduced mainly during sample preparation
and secondly by the level of sensitivity of the
instrument.

The major disagreement among researchers is over which
methods are most effective and how best to limit error.
The most common method of mercury analysis is cold vapor
atomic absorption spectrometry, which measures total
mercury; gas chromatography is generally accepted as the

most sensitive method for organomercurial determinations
at present. Controversy continues over the relative
value of such methods as cold vapor atomic absorption
spectroscopy and neutron activation analyses for total
mercury determinations.

The use of standard methods and environmental
reference samples within a laboratory reduces the
likelihood of systematic errors and improves the
accuracy of various analytical methods. Atomic
absorption equipment is calibrated with synthetic
standard samples, and some variabilities among operators
and instruments can be controlled by automation. The
final results can be checked by interlaboratory
comparisons. Results from analyses of blind samples
have been compared since the mid-1960s and the
discrepancies have decreased. See Appendix A for a
fuller discussion of analytical methods.

CHAPTER 1

THE GLOBAL CYCLE OF MERCURY

THE NATURAL MERCURY CYCLE

In considering sources, mechanisms of transport, and sinks of materials in the environment, it is often useful to consider their global cycle. Models can offer an initial insight into the general behavior of an element and often establish a framework for subsequent research. By considering both "pre-man" and present-day elemental global cycles, Garrels et al. (1973) have shown that these models also are helpful for identifying and evaluating anthropogenic impacts on fluxes between different reservoirs in nature.

Insights from Existing Models

Mercury has attracted more attention than many other trace elements in the past few years, and the recent literature contains a large body of information on its distribution in the environment. This information has facilitated the development of several global mercury models (Kothny 1973, Korringa and Hagel 1974, Wollast et al. 1975, Abramovskiy et al. 1975). Van Horn (U.S. EPA 1975b) has also developed a detailed mercury balance for the United States. Although details of calculations and basic assumptions vary among the different authors, several important conclusions have been drawn from these previous models about the behavior of mercury in nature. Some of the more conspicuous can be summarized as follows:

1. The atmosphere plays an important role in the mobilization of mercury. The earth surface-to-atmosphere flux is several times larger than that occurring between continents and oceans.
2. The earth surface-to-atmosphere flux seems to involve mainly elemental mercury vapor, whereas the flux between continents and oceans involves inorganic divalent mercury, much of it associated with dissolved

15

and particulate organic matter. In terms of quantity, alkylated forms of mercury do not contribute appreciably to the global mobilization of the element. A biological cycle that involves the transfer of alkylated forms of mercury does, however, exist. It is discussed in more detail below as well as in Chapters 3 and 4.

3. According to various estimates, the residence time for mercury in the atmosphere varies from 5.5 to 90 days, and the proportion of man-made mercury in the atmosphere ranges from 10 to 80 percent.

4. Use of mercury by man and subsequent emissions to land, rivers, and lakes, together with increased erosion rates, have elevated the mercury content of lakes and rivers by a factor of 2 to 4. Increases in total oceanic mercury concentrations have been negligible. The mercury content of soil appears to have increased by about 0.02 percent (Wollast et al. 1975).

Assumptions Involved in Existing Models

Estimates of the residence time of mercury as well as of the proportion of anthropogenic mercury in the atmosphere differ considerably. Therefore, it is worthwhile to consider some of the assumptions on which the published models are based.

Many calculations on the atmospheric flux of mercury are derived from measurements of dated Greenland ice cores. Recent evidence (Weiss et al. 1975) suggests that there has been no systematic change in the mercury concentration in these ice cores over the past 150 yr. In view of this evidence, previous observations of increased mercury concentrations in recently deposited snow and subsequent conclusions that this increase is related to anthropogenic inputs still appear to be unresolved. The mercury deposition at Greenland calculated from the 1975 Weiss et al. data seems, rather, to reflect polar and oceanic deposition rates not influenced by man.

Measurements of the natural mercury flux from the earth's surface to the atmosphere are mainly based on degassing rates measured in California. Degassing rates used by various authors range from 0.0014 to 10 $\mu g/m^2/day$. In addition, some investigators have assumed that degassing takes place only over the continents, whereas others consider this rate equal for continents and oceans. Further research is needed to clarify this issue, although Wollast et al. (1975) argue rather convincingly that the oceans must be a source for atmospheric mercury. The low atmospheric mercury concentrations over the oceans indicate, however, that lower degassing rates can be expected than those that occur on the continents.

Removal of mercury from the atmosphere has been thought to occur mainly via washout by precipitation. Conflicting conclusions about changes in the atmospheric mercury concentrations before and after rain events leaves the validity of this assumption in doubt. This topic is treated in further detail in the next chapter.

The assumption that mercury concentrations in air are constant with increasing altitude must also be questioned. Recent data by Abramovskiy et al. (1975) show that the mercury concentration decreases exponentially with increasing altitude. The decrease has been shown to follow the relationship $C_z = C_0 e^{-kz}$, where C_0 = ground concentration, C_z = concentration at any altitude, z, and k = a constant, approximately equal to 10^{-3} per meter. These considerations make it likely that previous estimates of the atmospheric mercury pool have been excessive.

An Updated Model

The preceding discussion covers some of the salient features of previous publications. Figures 1.1 and 1.2 combine data from these previous models with the new information to summarize current knowledge of the global mercury cycle in a pre-man and a present-day framework. Mercury fluxes and reservoirs have been calculated from the data presented in Tables 1.1 and 1.2 and from other sources discussed elsewhere in this report.

The pre-man cycle presented in Figure 1.1 is a modification of the model presented by Wollast et al. (1975). The most important assumptions used in constructing these models can be summarized as follows:

• The mercury deposition rate over polar regions, derived from Weiss et al. (1975), is taken to be 22 × 10^{-6} g/m²/yr. A similar deposition rate is assumed for oceanic areas. If the average mercury concentration in air for these areas is 0.7 × 10^{-9} g/m³ (see Chapter 2), the deposition velocity is approximately 10^{-3} m/s (wet + dry) (Chamberlain 1960). If the deposition velocity remains constant over oceanic shelf areas and continents, the rate of deposition for these two areas is (1.5/0.7) × 22 × 10^{-6} = 47 × 10^{-6} g/m²/yr and (4.0/0.7) × 22 × 10^{-6} = 126 × 10^{-6} g/m²/yr. The assumption that the deposition velocity for polar, open ocean, and oceanic shelf areas is similar seems quite reasonable, whereas the assumption that the deposition velocity over continents is the same remains to be tested.

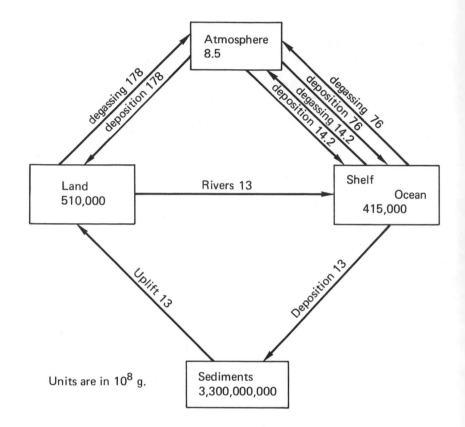

Units are in 10^8 g.

SOURCE: Modified from Wollast et al. (1975). Pages 145-166, *Ecological Toxicology Research: Effects of Heavy Metals and Organohalogen Compounds.* Reprinted with permission from Plenum Press, New York.

FIGURE 1.1   Pre-man global cycle for mercury.

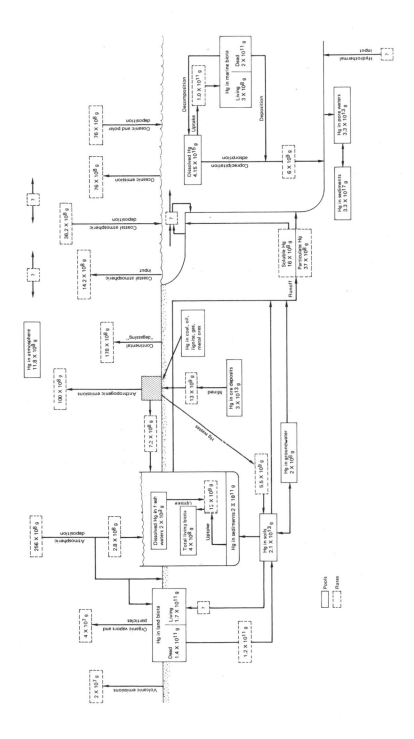

FIGURE 1.2 Present-day global cycle for mercury.

TABLE 1.1 Parameters Used in Constructing the Mercury Budgets

Area of the earth	510 $\times 10^{12}$ m^2
Area of the continents	149 $\times 10^{12}$ m^2
Area of polar regions	14.7 $\times 10^{12}$ m^2
Area of oceanic shelf regions	30 $\times 10^{12}$ m^2
Area of oceans	361 $\times 10^{12}$ m^2
Area of earth's freshwater systems	2.6 $\times 10^{12}$ m^2
Continental run-off volume	3.2 $\times 10^{16}$ ℓ/yr
Suspended sediment yield	183 $\times 10^{14}$ g/yr
Precipitation over continents	1.07 $\times 10^{17}$ ℓ/yr
Precipitation over oceans	4.11 $\times 10^{17}$ ℓ/yr
Atmospheric input of sea salt	1 $\times 10^{15}$ g/yr
Atmospheric input of volcanic dust	25 $\times 10^{12}$ g/yr
Atmospheric input of dust	5 $\times 10^{14}$ g/yr
Atmospheric input of vegetation	2 $\times 10^{14}$ g/yr
Open ocean primary productivity	16.3 $\times 10^{15}$ g C/yr
Coastal zone primary productivity	3.6 $\times 10^{15}$ g C/yr
Primary productivity in upwelling areas	0.1 $\times 10^{15}$ g C/yr
Open ocean fish production	1.6 $\times 10^{12}$ g/yr
Coastal fish production	1.2 $\times 10^{14}$ g/yr
Fish production in upwelling areas	1.2 $\times 10^{14}$ g/yr
Net primary production of land biota	47.6 $\times 10^{15}$ g/yr
Net primary production of lake and stream biota	0.6 $\times 10^{15}$ g/yr

C = Carbon

TABLE 1.2 Mercury Content of Global Reservoirs

Pool	Pool Mass (g)	Hg Concentration	Total Hg in Pool (g)
Atmosphere	5 $\times 10^{21}$		8.5 $\times 10^8$
Oceanic and polar		z_0 = 0.7 ng/m^3	2.4 $\times 10^8$
Oceanic shelf areas		z_0 = 1.5 ng/m^3	0.5 $\times 10^8$
Continental		z_0 = 4.0 ng/m^3	5.6 $\times 10^8$
Lithosphere			
Soils	3 $\times 10^{20}$	71 ng/g	2.1 $\times 10^{13}$
Freshwater sediments	6.5 $\times 10^{17}$	330 ng/g	2.1 $\times 10^{11}$
Oceanic sediments	10.2 $\times 10^{23}$	330 ng/g	3.3 $\times 10^{17}$
Hydrosphere			
Oceans	1.37 $\times 10^{24}$	0.03 ng/g	41.5 $\times 10^{14}$
Sediment porewaters	3.3 $\times 10^{23}$	0.1 ng/g	3.3 $\times 10^{13}$
Lakes and rivers	3 $\times 10^{19}$	0.06 ng/g	0.2 $\times 10^8$
Glaciers	2 $\times 10^{22}$	0.05 ng/g	0.1 $\times 10^{11}$
Groundwater	4 $\times 10^{18}$	0.05 ng/g	0.2 $\times 10^7$
Biosphere			
Living land biota	8.3 $\times 10^{17}$	200 ng/g	1.7 $\times 10^{11}$
Dead land biota	7 $\times 10^{17}$	200 ng/g	1.4 $\times 10^{11}$
Living marine biota	1.5 $\times 10^{15}$	200 ng/g	0.3 $\times 10^7$
Dead marine biota	10 $\times 10^{17}$	200 ng/g	0.2 $\times 10^{10}$
Living freshwater biota	2 $\times 10^{15}$	200 ng/g	0.4 $\times 10^7$

z_0 = surface altitude

● Using mercury degassing rates measured by McCarthy et al. (1969), Van Horn (U.S. EPA 1975b) conducted a state-by-state inventory of mineralized and nonmineralized areas. From the results, an average degassing rate of 130 $\times$ 10^{-6} g Hg/m^2/yr was calculated for the United States. This rate was assumed to remain the same for all continents. The degassing rates that support the atmospheric mercury concentrations for oceanic shelf, open ocean, and polar regions were then calculated to be (1.5/4.0) $\times$ 130 $\times$ 10^{-6} = 49 $\times$ 10^{-6} g/m^2/yr and (0.7/4.0) $\times$ 130 $\times$ 10^{-6} = 23 $\times$ 10^{-6} g/m^2/yr, respectively. These calculations balance the input rates quite well.

● Since no increase in mercury deposition rates has been observed in polar regions, it was assumed that emissions of mercury to the atmosphere from anthropogenic sources would result in increased deposition rates only over continents and oceanic shelf areas. As additional data become available this assumption will undoubtedly have to be revised since atmospheric constituents in the Northern Hemisphere are transported from east to west over considerable distances. From an environmental protection point of view, however, this assumption is conservative because it maximizes mercury deposition rates for continental and oceanic shelf areas.

● Since there are no data to indicate that degassing rates in oceanic shelf and open ocean areas have increased over the past century, it was assumed that these two areas realized a net increase of 22 $\times$ 10^8 g/yr from the atmosphere (see Figure 1.2).

● Flux rates between continental, oceanic shelf, and open ocean air masses and between shelf and open ocean water masses cannot be evaluated with current information. Evidence suggests, however, that much of the riverborne mercury is deposited initially in estuarine and continental shelf areas.

Many conclusions derived from the two models just presented are similar to those discussed earlier, but there are some differences as well. The residence times of mercury for the various reservoirs calculated here are: soils, 1000 years; atmosphere, 11 days; oceans, 3200 years; and sediments, 2.5 $\times$ 10^8 years. The shorter residence time for atmospheric mercury calculated in this report (see Chapter 2) is at variance with results from several previous models. The discrepancy can only be resolved with additional measurements both for different altitudes and for remote areas. Additional data on atmospheric mercury levels in remote areas would be especially helpful since the degree of variation of tropospheric gases has been shown to be inversely

proportional to their residence time in a model proposed
by Junge (1974).

The model further indicates that from 25 to 30 percent
of the atmospheric mercury burden is due to man-made
emissions. It is still possible that a fraction of the
atmospherically deposited mercury is again
revolatilized. The magnitude of this
deposition/volatilization phenomenon is presently
unknown. Similarly, the mercury burden of rivers (water
plus bottom and suspended sediments) has increased by a
factor of 4 when compared with pre-man levels. This
increase is presumably due to greater suspended sediment
loads as well as to use of mercury by man.

Because of large variations in the mercury levels of
freshwater and land biota and lack of historical data
for a wide range of biological specimens, the models
fail to resolve the question of whether increases have
occurred in these compartments. The evidence for
mercury increases in fish is discussed in Chapters 4 and
5. However, a survey taken for this study of mercury
concentrations in sediment cores from freshwater lakes
and estuaries indicates that these reservoirs have
levels approximately 2 to 5 times higher than in
precultural times (Shimp et al. 1970, Lindberg et al.
1975, U.K. Department of the Environment 1976).

THE GLOBAL CYCLE OF METHYLMERCURY

Chapter 3 reviews the mechanisms in the biological
cycling of mercury. The discussion in this section,
consequently, is confined to alkylated mercury flux
calculations for fresh and ocean water systems.

Although many laboratory measurements of methylation
and demethylation rates have appeared in the literature
(Booer 1944, Kimura and Miller 1964, Wood et al. 1968,
Jensen and Jernelov 1969, Landner 1971, Yamada and
Tonomura 1972, Bisogni and Lawrence 1973, Spangler et
al. 1973a, Billen 1973), quantifications of the flux of
alkylated mercury for in situ natural conditions are
lacking. Methylation rates in natural freshwater
sediments spiked with inorganic mercury range from about
5 to 15 percent per year per gram of sediment. Although
natural bacterial populations were maintained (i.e.,
there was no microbial enrichment), these experiments
reflect only extremely polluted areas (100 μg Hg/g).
Demethylation rates have also been measured by Spangler
et al. (1973a) and by Billen (1973), and they were in
the same range as the methylation rates in the mercury-
contaminated samples. A delicate balance between
methylation and demethylation is thus implied by these
results.

These rates, however, do not reflect more commonly encountered mercury concentrations; nor do they necessarily represent the different redox, pH, and microbial variations in nature. Billen and Wollast (1973) have concluded that the most intense mercury methylation occurs in the transitional "oxidizing-anaerobic" zones of natural sediments. In sediments, pH commonly ranges between 6 and 8, while E_h ranges from approximately -0.2 to +0.4 V for these same zones. Bacterial demethylation seems to occur over a broader range of redox conditions. These physicochemical and microbiological factors have made in situ field measurements of methylation and demethylation rates difficult. In addition, extremely fast methylmercury cell diffusion rates and analytical problems make it necessary to calculate alkylmercury fluxes by indirect methods.

Based on the flux rates presented in Figures 1.1 and 1.2, it is possible to make some crude approximations of the net methylation rates in fresh and oceanic waters. The uptake rate can be calculated if the fish population is assumed to be the ultimate sink for methylmercury in these two reservoirs and that once methylmercury is produced, it is taken up by fish and other biota. According to Wetzel (1975), the annual fish production in fresh waters ranges from about 1 to 100 $g/m^2/yr$. If one takes 20 $g/m^2/yr$ as an annual average, the global fish production for these waters is 52×10^{12} g/yr. If one further assumes that the average mercury content of fish is 200 ng/g and that all of this is methylmercury, the annual uptake rate in the world's freshwater systems is 10^7 g methylmercury/yr. With the same basic assumptions and a total fish production rate of 2.4×10^{14} g/yr (Ryther 1969), the annual methylmercury uptake by oceanic fish is 4.8×10^7 g. On the assumption that the methylmercury uptake rate in fish is approximately equal to the net methylation rate of each reservoir, the rate would be 10^7 g/yr and 4.8×10^7 g/yr for fresh waters and oceans, respectively. Since open ocean fish production represents only 1 percent of the total oceanic production, it appears that most of the methylation is confined to nearshore and upwelling areas. The actual sites of methylation remain to be investigated.

If it is assumed that methylation takes place in bottom sediments in both reservoirs and is restricted to the upper 5 to 10 cm (Jernelov 1970), it is also possible to calculate the fraction of total mercury that is available for methylation. The amount of total mercury in the upper 10 cm of freshwater and oceanic sediments is approximately 2×10^{11} g and 3×10^{13} g, respectively (Kothny 1973). The average net methylation rate for the two reservoirs is thus less than 0.005 percent per year. If the oceanic methylation site is restricted primarily to estuarine and continental shelf sediments, this rate is increased by approximately an order of magnitude.

Although the calculations above contain many simplifying assumptions (for instance, ignoring diffusion rates of methylmercury from sediment to water and the fact that biota other than fish are the ultimate sink), they should nevertheless reflect order-of-magnitude fluxes. The model explains the extremely low alkylmercury concentrations in water and sediments (Chau and Saitoh 1973, Andren and Harriss 1975). These calculations also indicate that the biological cycle is delicately balanced and that small perturbations in input rates and the chemical form of mercury can result in increased methylation rates in sensitive ecosystems. It is also apparent that studies of the behavior of methylmercury in nature must be intimately linked with bioenergetic studies.

MAN-MADE MERCURY EMISSIONS IN THE UNITED STATES

The most recent detailed analysis of man-made versus natural mercury emissions in the United States is that of Van Horn (U.S. EPA 1975b). The analysis includes a detailed summary of past, present, and future mercury consumption patterns in the coterminous United States. This information is presented in Table 1.3. Total mercury losses to air, water, and land by different activities on a national basis are presented in Table 1.4. In addition to consumption through commercial, industrial, and consumer uses, which alone accounts for more than 60 percent of the total man-related mercury losses, the other major sources of anthropogenic mercury emissions are mining and related activities, energy-related activities, manufacturing, and processing.

Data from Table 1.4 indicate that the total anthropogenic mercury losses in the United States for 1973 were approximately 1.5×10^9 g. Of these, 31 percent was emitted to air, 6 percent to water, 63 percent to land, and 1.5×10^8 g were recycled. The natural degassing rate for the United States was estimated to be about 1×10^9 g/yr. If it is assumed that mercury emitted from the various sources has the same atmospheric residence time, it is evident that 32 percent of the atmospheric mercury burden in the United States is anthropogenic in origin.

A state-by-state estimate of natural and man-made mercury emissions to air and water is presented in Table 1.5. The man-made increases in mercury discharges to

TABLE 1.3 Mercury Consumption in the United States, 1970-1973 and Projected for 1985

End Use Activity	1970	1971	1972	1973	Projected for 1985
			$(10^6$ g)		
Agriculture	62.3	50.8	63.2	62.9	24.1
Catalysts	77.0	34.3	27.5	23.2	17.2
Dental preparations	78.6	64.4	102.6	92.2	129.0
Electrical apparatus	548.7	572.6	535.0	619.2	662.0
Caustic chlorine	516.4	421.5	396.2	449.6	481.6
Laboratories	62.1	46.7	20.4	22.6	?
Industrial instruments	166.2	134.4	225.0	246.1	447.2
Paints	355.9	296.0	282.8	261.5	106.6
Pharmaceuticals	23.7	23.0	19.9	20.8	17.2
Other	209.3	170.0	147.2	69.1	206.4
TOTAL (rounded)	2100	1814	1820	1867	2091

SOURCE: Derived from U.S. EPA (1975b).

TABLE 1.4 Total Mercury Losses in 1973 for the Coterminous United States

Activity Description	Total Losses to			Total Hg Lost	Total Hg Recycled
	Air	Water	Land		
			$(10^6$ g)		
Hg mining, smelting and processing	7.85	0	0.42	8.27	0
Other mining activities	50.71	3.07	4.91	58.09	0
Energy related activities	104.5	2.18	42.07	148.73	0
Manufacturing and processing (see Table 1.3)	26	21.72	271.33	319.05	20.66
Commercial, industrial and consumer consumption	282.2	60.73	646	990	126.11
TOTAL	471.26	87.70	964.73	1524.73	146.77
Sewage	4.01	19.92	22.88	46.0	
Natural degassing	1018.7			1018.7	

SOURCE: Derived from U.S. EPA (1975b).

TABLE 1.5 Mercury Emissions in the United States, 1973

	Man-made Losses to		Natural Losses to		Percent Due to Man	
	Air	Water	Air	Water	Air	Water
			$(10^6$ g/yr)			
Alabama	9.9	2.5	9.6	4.1	51	38
Arizona	22.3	2.4	85.8	11.3	21	18
Arkansas	3.0	1.8	9.8	2.0	23	47
California	36.8	14.2	118.2	26.3	24	35
Colorado	3.5	1.6	39.2	6.0	8	21
Connecticut	6.0	2.0	1.8	0.01	77	99
Delaware	1.9	0.5	0.4	0.01	83	98
Florida	12.1	4.6	10.2	0.8	54	85
Georgia	9.5	3.2	22.0	2.7	30	54
Idaho	2.8	0.6	31.3	6.3	8	9
Illinois	22.4	8.0	10.5	1.4	68	85
Indiana	12.2	3.8	6.8	3.2	72	54
Iowa	5.6	2.0	10.6	5.9	35	25
Kansas	3.7	2.1	15.5	2.2	19	49
Kentucky	10.5	2.1	7.5	1.2	58	64
Louisiana	8.4	2.6	8.5	0.5	50	84
Maine	2.5	0.7	5.8	0.3	30	70
Maryland	7.6	3.0	3.7	0.1	67	97
Massachusetts	11.4	3.9	3.0	0.6	79	87
Michigan	18.8	5.2	10.7	0.8	64	87
Minnesota	7.0	2.7	14.9	6.0	32	31
Mississippi	5.1	1.4	8.9	4.3	34	25
Missouri	13.8	3.4	13.0	4.6	50	43
Montana	4.0	1.0	55.0	1.9	7	34
Nebraska	2.3	0.7	14.5	1.9	14	27
Nevada	5.6	0.3	83.1	4.2	6	7
New Hampshire	1.4	0.3	3.4	0.1	29	75
New Jersey	14.6	3.4	1.4	0.2	91	94
New Mexico	4.9	0.5	45.9	20.7	10	2
New York	35.5	7.1	18.1	0.6	66	92
North Carolina	12.2	2.2	18.5	1.2	40	65
North Dakota	0.9	1.0	13.1	0.9	6	53
Ohio	25.3	4.5	7.8	1.8	76	71
Oklahoma	4.6	1.3	13.0	25.5	26	5
Oregon	3.6	0.8	72.7	2.6	5	24
Pennsylvania	34.8	5.1	17.0	0.8	67	86
Rhode Island	1.8	0.4	0.2	0.02	90	95
South Carolina	5.5	1.2	5.7	1.1	50	52
South Dakota	1.1	0.4	14.4	1.3	7	24
Tennessee	10.0	2.3	7.8	5.2	56	31
Texas	25.9	4.7	49.6	18.5	34	20
Utah	7.8	0.7	31.0	3.0	20	19
Vermont	0.7	0.2	2.2	0.2	24	50
Virginia	8.1	1.8	15.0	1.1	35	62
Washington	9.0	1.5	25.2	0.9	26	63
West Virginia	5.4	0.8	9.1	0.6	37	57
Wisconsin	8.4	2.3	10.3	0.9	45	72
Wyoming	0.7	0.1	36.8	2.9	2	3

SOURCE: Derived from U.S. EPA (1975b).

air must be interpreted with care because each state has a different natural degassing rate. Thus, states with a low natural rate might show a greater increase even though the man-made emissions are lower than those in a state with high emission rates from both sources. These calculations can only be validated through a properly administered monitoring program. Calculations of man-made mercury emissions to water indicate that annual natural emissions have at least been doubled by anthropogenic discharges in Connecticut, Delaware, Florida, Illinois, Louisiana, Maine, Maryland, Massachusetts, Michigan, New Hampshire, New Jersey, New York, Ohio, Pennsylvania, Rhode Island, and Wisconsin.

Because the relation between mercury emissions and population density is an important consideration, Van Horn expressed the data in Table 1.5 on the basis of population. The mean mercury emissions in kilograms per 1000 population per year, calculated for the coterminous United States in 1973, were: air, 1.81 ± 0.53; water 0.41 ± 0.08; land, 4.55 ± 1.20. Table 1.6 identifies so called "hot spots" in terms of mercury emissions to air, water, and land as a function of population density. These "hot spots" are states with exceptionally high levels of mercury emissions per 1000 population and have been related to either chloralkali plants or copper smelting facilities. On the other hand, Van Horn's analysis showed that the presence of chloralkali plants and copper smelting facilities does not necessarily lead to elevated concentrations.

It is not yet clear how valuable these data are for projecting the mercury burden of aquatic and terrestrial biota and the human population, but their value is apparent in establishing monitoring and health evaluation programs. Since the behavior of mercury is very much a function of its chemical form, potential biomethylation "hot spots" might be predicted by evaluating the data presented in Tables 1.5 and 1.6 together with prevailing physical, geological, and chemical regimes for each region. For the aquatic system, various factors must be considered such as pH, redox conditions, alkalinity, buffering capacity, suspended sediment load, and lake or river geomorphology.

Within this context, Brouzes et al. (1977) have found that poorly buffered lakes in the Canadian shield region are particularly sensitive to the impact of acid rain. Fish collected in these lakes exhibit high mercury concentrations even when the lakes and rivers are not near industrial activities, roads, or villages. Similarly sensitive areas should be identified in the United States in addition to systems that have received direct mercury spills. The release of mercury to the

TABLE 1.6 Annual Man-Made Mercury Emissions in the Coterminous
United States, 1973

Level of Emissions	Air	Water	Land
		(g/1000 pop.)	
"Hot spots"	Arizona, Montana, Nevada, New Mexico, Utah $> 4 \times 10^3$	Arizona, Montana $> 0.8 \times 10^3$	Delaware, Kentucky, Louisiana, West Virginia $> 9 \times 10^3$
Moderately elevated	Georgia, Idaho, Kentucky, Missouri, Pennsylvania, West Virginia $(2.3\text{-}4) \times 10^3$	Delaware, Idaho, Kansas, Nevada, South Dakota, Tennessee, Utah $(0.49\text{-}0.8) \times 10^3$	Georgia, Maine, Montana, New Jersey, Tennessee, Washington $(5.7\text{-}9) \times 10^3$
Below national average	Colorado, Kansas, Nebraska, Vermont $< 1.28 \times 10^3$	Florida, Vermont $< 0.33 \times 10^3$	Arizona, Florida $< 3.4 \times 10^3$

SOURCE: Adapted from U.S. EPA (1975b).

environment from coal-fired power plants, mining, and smelting operations is of special concern since these sources are currently uncontrolled and the use of coal is expected to increase significantly.

CHAPTER 2

FORMS AND OCCURRENCE OF MERCURY IN THE ENVIRONMENT

Because different chemical forms of mercury possess different reactivity and toxic characteristics, it is important to identify these in the environment. Two major classes of mercury compounds are generally distinguished, inorganic and organic. From the standpoint of risk to human health from environmental exposures, the alkylated mercurials within the class of organic mercury compounds are the most important because of their toxic properties and their tendency to bioaccumulate. Of the alkylmercurials, methylmercury presents the greatest risk. However, both the inorganic forms (such as metallic mercury and mercury sulfide, the principle mercury ore) and the organic forms of mercury are subject to conversion in the environment, and thus, their availability is important.

MERCURY IN THE ATMOSPHERE

The sources of mercury to the atmosphere have not been completely identified, quantified, or characterized. Our present understanding, however, indicates that the most important sources are rock and soil volatilization, volcanic exhalations, biological emanations, suspended continental dust, and various anthropogenic inputs. On the basis of data obtained by analyzing dated Greenland and Antarctic snow cores (Weiss et al. 1971), several authors (Korringa and Hagel 1974, Heindryckx et al. 1974, Wollast et al. 1975) have suggested that the annual natural flux of mercury to the atmosphere is from 25,000 to 30,000 metric tons per year (1 ton = 10^6 g). This includes the input from both terrestrial and oceanic sources, although the relative importance of the two is not known. By the same analogy, these authors argue that present-day input of mercury to the atmosphere from both natural and anthropogenic sources ranges from 41,000 to 50,000 metric tons, because data from the snow cores showed this increase in the more recent deposits. As shown below, recent measurements at

another station in Greenland do not show this decrease
with depth in the snow cores. The atmospheric cycle of
mercury must be analyzed by other means, as has already
been discussed in Chapter 1.

For a firmer understanding of present knowledge about
its atmospheric cycle, a compilation of recently
published mercury levels discussed in this section is
presented in Table 2.1. Because relatively few
measurements have been made of large-scale
geographically representative areas, it is prudent to
use observed ranges rather than to emphasize mean or
average concentrations. Since the latter values are,
nevertheless, quite useful for a better understanding of
the mercury cycle, they are included. Remote oceanic
areas have atmospheric mercury concentrations of around
0.7 ng/m³, almost all in the vapor form. Remote rural
areas exhibit mercury concentrations that cluster around
4 ng/m³, with less than 5 percent in particulate form.
Urban areas are quite variable, but seem to exhibit
concentrations of somewhat less than 10 ng/m³ with
variable particulate mercury fractions. Point sources
such as volcanoes, mines, and industries can increase
the atmospheric burden considerably, and Carr and
Wilkniss (1973) have argued that volcanic emissions can
account for any variability of the mercury burden in
snow cores.

Mercury Vapor

Although by no means substantiated on a global scale,
many investigators have hypothesized that elemental
mercury, Hg⁰, is the major form of mercury vapor
(Williston 1968, McCarthy et al. 1969, Johnson and
Braman 1974). This should be especially true near
geothermal and volcanic areas. Because of
disproportionation reactions between mercuric oxide and
metallic mercury at high temperatures, it is also
thought that more than 97 percent of the mercury emitted
from coal-fired steam plants is in the form of elemental
mercury (Billings et al. 1973, Klein et al. 1975).

Johnson and Braman (1974) recently developed a
technique for distinguishing between different chemical
forms of mercury vapor in air. Air is first drawn
through a glass fiber filter and then through a series
of selective adsorption tubes that contain HCl-treated
siliconized Chromosorb W, NaOH-treated Chromosorb W,
silver-coated glass beads, and gold-coated glass beads.
This system collects particulate mercury, mercury(II)-
type compounds, methylmercury-type compounds, metallic
mercury, and dimethylmercury. The measurements
substantiate previous observations that mercury in air

TABLE 2.1   Summary of Current Data on Atmospheric Mercury Levels for
Various Locations

	Range	Mean
	(ng/m^3)	
A. Remote and Rural Areas		
Oceanic		
Particulate	< 0.005–0.06	<0.15
Vapor	0.6–0.7	0.7
Non-mineralized terrestrial		
Particulate	< 0.005–1.9	0.15
Vapor	1–10	4.0
Volcanic		
Particulate + vapor	20–37,000	–
Mineralized terrestrial		
Particulate + vapor	7–20,000	–
B. Urban Areas		
Particulate	< 0.01–220	2.4
Vapor	0.5–50	7.0
C. Industrial*		
Vapor	7–5,000,000	–

*These measurements include chlor-alkali plants, thermometer factories, smelters, and
mercury mines.

is primarily a vapor. Concentration ranges and means
for 11 stations around Tampa, Florida, are shown in
Table 2.2 (Johnson and Braman 1974). Their data
indicate that dimethylmercury is rarely encountered in
the atmosphere, but that other forms are present in
varying amounts. The particulate fraction averaged 4
percent of the total mercury collected. The vapor
species were averaged as follows: Hg(II)-type, 25
percent; methylmercury-types, 21 percent; Hg^0, 49
percent; and dimethylmercury, 1 percent. The authors
pointed out that Hg^0 and methylmercury-type vapors
exhibited fairly uniform concentrations whereas the
other forms varied considerably and the particulate
mercury was strongly correlated with wind conditions.
The authors indicated that methylated mercury was
emitted from the terrestrial system as well as from a
highly polluted bay nearby.

These and all previous data must, however, await
further evaluation and validation. The different
species of mercury collected by the previously described
technique are essentially defined operationally, i.e.,
glass fiber filters pass a sizable number of particles
of less than 0.3 μm; and it is therefore possible that a
portion of the first tube collects particulate matter as
well. This problem will be present whenever filtration
techniques are used to separate particulate from vapor
phase mercury. Disproportionation reactions also
readily take place, especially when in contact with many
different collection surfaces. Extremely difficult
analytical techniques as well as limited geographical
sampling do not permit further conclusions as to the
mercury speciation in air. More emphasis must be placed
on obtaining an atmospheric mercury data base from
culturally and geographically representative areas of
the world.

Mercury in Rain and Snow

Even less information exists on mercury in rain and
snow, and it is presently difficult to predict levels
for various locations. The published data are
summarized in Table 2.3. A partial explanation for this
lack of data can be attributed to difficulties
encountered in rain sampling as well as in the
interpretation of the data. Except for areas of direct
input, most measurements show levels of from 0.01 to 0.1
μg/l. Some contradictions seem to exist in the
literature as to the rainout and washout efficiencies of
mercury. Measurements by McCarthy et al. (1969) showed
that a heavy rainstorm very efficiently removed mercury
from air whereas Johnson and Braman (1974) reported

TABLE 2.2 Levels and Chemical Forms of Mercury in Tampa, Florida[a]

| | ng/m^3 | | | | | |
	[Hg] part.	[Hg(II)] vap.	[CH$_3$Hg]	[Hg^O]	[CH$_3$HgCH$_3$]	Σ Hg
Range	<0.03–13	<0.03–220	<0.07–119	<0.03–49	<0.03–3	1.8–298
Mean[b]						
Day	0.27	0.86	0.63	2.67	0.05	4.48
Night	0.17	1.58	1.56	5.03	0.06	8.40

[a]From 11 stations around Tampa, Florida at two different dates.
[b]Mean values collected at one station for 33 consecutive 2 hour samples.

SOURCE: Johnson and Braman (1974). Reprinted with permission from Environmental Science and Technology 8:1003-1009. Copyright by the American Chemical Society.

TABLE 2.3 Mercury Levels in Precipitation

Location and Type	Range (μg/ℓ)
A. Rural U.K.	<0.2
Germany (1934)	0.05–0.48
Oak Ridge, Tenn., U.S.A.	<0.05–0.54
B. Snow Layers	
Sweden: surface layer	0.07
sublayer	0.21
near chlor-alkali plant	<2–11
Canada: urban	3.5–4
rural	<0.01–0.52
C. Greenland and Antarctica	
Before 1900*	0.013–0.169
After 1900*	0.040–0.23

*No apparent trends with age of snow deposit (Weiss et al. 1975).

essentially the same atmospheric mercury concentration before, during, and after a thunderstorm. Therefore, it is possible that mercury can have a significant dry removal component (vapor impact) as well. Until this question has been resolved use of historical records of snow cores to deduce the mercury content of snow at the time it was deposited is in doubt. It is entirely possible that the measured mercury concentrations in snow reflect variable proportions of dry plus wet input.

The chemical form of mercury in precipitation are not known, and attempts to determine the physical form of mercury in precipitation have met with limited success. This is mainly due to extremely difficult sampling problems. Mini dust storms often precede precipitation events. Since most investigators use automatic rain-collecting devices that collect time-integrated samples, avoiding this problem is difficult. Several reports on concentrations in precipitation include mercury added to rainwater by windblown dust of a very localized nature (i.e., from the immediate vicinity of a sampler). To separate dissolved and particulate mercury requires operationally defined procedures such as filtration or centrifugation, which also have inherent problems. The sample must also be properly preserved if losses to the walls or by volatilization are to be prevented.

Many of the severe sampling and analytical problems indicated above stem from the need to use operational definitions, i.e., based on collection techniques, when assigning proper values to the chemical and physical forms of the sampled mercury. Additionally, because recent data indicate no systematic change in mercury concentrations of snow cores from Greenland, the previous calculations of man's influence on the global mercury cycle need to be reevaluated.

MERCURY IN THE LITHOSPHERE

Mercury in the lithosphere has been reviewed in two recent reports (World Health Organization 1976b, U.K. Department of the Environment 1976). In mineralized areas, in some rocks, or where anoxic conditions exist, much of the mercury is in the form of its sulfide (cinnabar). This is the most insoluble form of mercury. However, cinnabar is mined for industrial and human consumption uses and when released into the environment is transformed into other mercury compounds, making it more readily available for potential methylation. Once mercury is thus mobilized, if it is not then recycled but, rather, released into the environment, conversion to its original geochemical form (for example mercury sulfide) and burial would be the best safeguard against transformation into methylmercury.

Statistical correlations between total sediment sulfur and mercury have been interpreted as evidence that mercury often occurs either as an insoluble sulfide or adsorbed onto the surface of sulfide minerals (mainly FeS_2). According to Andersson (1967) and Keckes and Miettinen (1970), mercury is predominantly bound to soils and sediments through association with organic matter. In addition, Vernet and Thomas (1972) and Thomas (1973) suggest that observed relationships between total mercury, iron, and phosphorus indicate that mercury is bound to an inorganic iron-phosphate complex, probably adsorbed onto hydrated oxide and manganese coatings of clay particles.

Divalent inorganic mercury can undergo reduction to elemental mercury. Certain bacteria of the genus Pseudomonas can perform the conversion (Magos et al. 1964, Furukawa et al. 1969). Processes leading to the formation of elemental mercury in soils and sediments are not well understood, but must form the basis for soil degassing of mercury. According to the WHO report (1976b:50): "Unfortunately, other than very crude generalizations, little is known of the details of kinetics of these processes in nature."

Within the last few years an enormous quantity of data on relative mercury concentrations in soils and sediments has appeared in the literature. Table 2.4 summarizes mercury levels in rocks, soils, and sediments from a recent review (U.K. Department of the Environment 1976). Mercury concentrations of unpolluted, nonmineralized soils can vary by 2 orders of magnitude with the lower concentrations at about 0.01 μg/g. The mean values for American and British soils were 0.071 μg/g (Shacklette et al. 1971) and 0.060 μg/g, respectively. The former value is most often used as the overall mean crustal concentration of mercury. Although highly variable, the mercury content of soils in mineralized areas can exceed 500 μg/g. Cinnabar ore deposits usually contain from 0.5 to 1.2 percent mercury. The mercury content of nonmineralized freshwater sediments exhibits ranges similar to those of soils, although the available data indicate a mean of about 0.3 μg/g. Concentrations of 800 to 1000 μg/g have been observed in polluted sediments, usually in close proximity to chloralkali plants. Aston et al. (1972) found a mean mercury value of 0.41 μg/g for North Atlantic sediments, and Weiss et al. (1972) found values ranging from 0.012 to 0.173 μg/g in sediments collected in the Pacific Ocean off the west coast of Mexico. Many more measurements of mercury in oceanic sediments are required to arrive at an average value although 0.33

TABLE 2.4 Mercury Concentrations in the Lithosphere

	Range	Mean
	(μg/g)	
Soils		
Sweden	0.01–1.0	0.070
Finland	0.02–0.2	0.060
England	0.01–15.0	0.060
Scotland	0.01–1.96	0.080
Japan	0.18–0.33	0.28
United States	0.01–4.7	0.071
Freshwater Sediments		
Sweden	0.034–26.5	0.3
Finland	0.05–170	–
England	0.01–1.026	–
United States	0.01–1200	0.3
Estuarine Sediments		
England	0.01–150	0.40
Canada	0.02–26.0	–
United States	0.01–0.5	0.33
Marine Sediments		
North Atlantic	–	0.41
Off West Mexico	0.012–0.173	–

SOURCE: U.K. Department of the Environment (1976). Environmental Mercury and Man, Pollution Paper No. 10. Reprinted with permission of the Controller of Her Britannic Majesty's Stationery Office.

$\mu g/g$ has been used in this and other reports for budget calculations (Wollast et al. 1975). According to Joensuu (personal communication, Rosensteil School of Marine and Atmospheric Sciences, University of Miami, 1977), his unpublished data suggest that this value should be revised to between 0.05 and 0.1 $\mu g/g$.

MERCURY IN NATURAL WATERS

Our knowledge of the chemical forms of mercury in natural waters is, at present, incomplete. The most common theoretical approach has been to predict the chemical forms of mercury by using thermodynamic calculations (Anfalt et al. 1968, Dyrssen and Wedborg 1974, Hem 1970, Wollast et al. 1975). These calculations predict that redox, pH, and ligand conditions are very important. Hem (1970) and Wollast et al. (1975) predict that in well-oxygenated waters (E_h > 0.4 V) dissolved mercury should exist as mercury(II)-type compounds, in moderately oxidizing to mildly reducing conditions (E_h -0.2 to 0.4 V) as Hg^0 or $Hg(II)$, and in reducing conditions (<-0.2) as Hg^0 or $HgS_2^=$. Anfalt et al. (1968) predicted that $Hg(II)$ would exist as $Hg(OH)_2$, $HgCl_2$, or $HgOHCl$ in fresh waters with a strong dependence on pH and pCl. Dyrssen and Wedborg (1974) also calculate that the most important species in seawater appear to be $HgCl_4^=$, $HgCl_3Br^=$, $HgCl_3^-$, $HgCl_2Br^-$, and $HgCl_2^0$ at relative concentrations of 65, 12, 12, 4, and 3 percent, respectively. Wollast et al. (1975) have shown that Hg_2^{++} can only exist at concentrations greater than 450 mg/l of total Hg, a level that is unlikely in natural waters.

There are, however, many kinetic factors that can change the speciation of mercury in the aquatic environment. These include: associations with dissolved organic matter, associations with suspended particulate matter (clays, hydrous oxides, and detrital organic matter), and methylation-demethylation processes. Fitzgerald and Lyons (1973) concluded that 50 to 60 percent of dissolved Hg in coastal waters could exist in association with organic matter of unspecified composition. Lindberg and Harriss (1973) and Andren and Harriss (1975) observed a strong association between dissolved organic matter and mercury in fresh, estuarine, and interstitial waters. The latter authors also suggested that substantial amounts of Hg are removed by flocculating organic matter in the transition zone between fresh and saltwater. Lindberg et al. (1975) have indicated that strong mercury-organic matter associations also prevent the highly insoluble mercury sulfide from precipitating in reducing sediments. This again emphasizes the importance of organic matter in regulating the behavior of mercury. The nature of this association is not at present well understood although functionalities such as -SH, -COOH, and -N are assumed to be responsible.

Because of its reactivity, mercury is strongly bound to suspended particulate matter, so that in most fresh water (especially in contaminated areas) more than 50 percent is transported in this manner (Cranston and Buckley 1972, Lindberg et al. 1975). Wollast et al. (1975) also report that hydrous metal oxides significantly reduce and modify the form of Hg in natural waters. These and other reports indicate that the concentration of dissolved Hg in fresh water varies between 0.02 and 0.06 $\mu g/l$, whereas typical oceanic values range from 0.01 to 0.03 $\mu g/l$. Data are scarce on the partitioning between dissolved and particulate phases in sea water, although they indicate that most of the Hg exists in the dissolved form (Carr et al. 1972).

Very little information has been presented on the levels of methylated mercury in fresh and marine waters although it has been shown that methylation takes place in fresh and coastal water sediments. Foreback (1973) indicated that approximately 50 percent of the dissolved Hg in a polluted Florida bay was methyl or phenylmercury. Chau and Saitoh (1973) measured methylmercury levels in several Canadian lakes. With a detection limit of 0.24 ng/l they found no methylmercury in unfiltered Great Lakes water and a concentration of 0.5 to 0.7 ng/l in four smaller, polluted lakes. Andren and Harriss (1975) could detect no methylmercury in samples from rivers and coastal waters of the eastern Gulf of Mexico. It thus appears that where methylmercury compounds are formed they are rapidly taken up by biota, and that nonpolluted waters consequently contain less than 0.2 to 1.0 ng/l of Hg in the methylated form.

Field and laboratory measurements nevertheless suggest that, even with the extremely low concentrations of methylated forms of mercury present in the aquatic environment, the major form of mercury in the biota is still methylated. This indicates that uptake rates and subsequent bioaccumulation of mercury must be rapid. It is, therefore, imperative that the natural methylation-demethylation balance not be disturbed by the discharge of mercury (in whatever form) and excess nutrients to the environment.

CHAPTER 3

CHEMICAL AND BIOCHEMICAL MECHANISMS FOR
METHYLATION AND DEMETHYLATION; KINETICS

"The case of mercury pollution has clearly
demonstrated the profound importance of
understanding biologically-mediated transformation
reactions that yield metal-organic compounds with a
high potential for bioaccumulation and toxicity."
(NRC 1977)

BIOMETHYLATION OF MERCURY

Three methylating coenzymes have been discovered that
participate in biomethylation reactions in biological
systems: (1) S-adenosylmethionine, (2) N^5-
methyltetrahydrofolate, and (3) methyl-B_{12}. Both from a
theoretical viewpoint and from direct experimental
evidence, it has been established that only methyl-B_{12}
is capable of methylating soluble inorganic mercury
salts to methylmercury and dimethylmercury. This
reaction occurs both under anaerobic and aerobic
conditions (Wood et al. 1968; Hill et al. 1970; Adin and
Espenson 1971; Neujahr and Bertilsson 1971; Schrauzer et
al. 1971, 1973; Wood 1971). Recent studies have
elucidated detailed mechanisms for the biosynthesis of
methylmercury and dimethylmercury from methyl-B_{12} (Imura
et al. 1971; DeSimone et al. 1973; Wood 1973, 1974,
1975a, 1975b). The second-order rate constant for the
biosynthesis of methylmercury from mercuric acetate has
been determined as $3.7 \times 10^2 \ s^{-1} \ m^{-1}$; and, therefore,
under optimum conditions the kinetics for methylmercury
synthesis have been shown to be extremely rapid
(DeSimone et al. 1973). The rate at which methylmercury
is formed in any environment is largely determined by
the available concentrations of soluble inorganic
mercuric ion and methyl-B_{12} (DeSimone et al. 1973, Wood
1975a).

The biosynthesis of methylmercury in sediments and by
bacteria isolated from sediments is well established
(Jensen and Jernelov 1968; Langley 1971; Bisogni and
Lawrence 1973, 1975a, 1975b; Vonk and Sijpesteijn 1973;

Spangler et al. 1973b; Jernelov 1974). It has also been confirmed that methylmercury synthesis is catalyzed by bacteria which have B_{12} dependence (Vonk and Sijpesteijn 1973). The rate of methylmercury biosynthesis can be enhanced by adding B_{12} to certain bacterial cultures (Vonk and Sijpesteijn 1973).

The biosynthesis of methylmercury in fungi is less clear, especially since it has been reported that Neurospora (a B_{12} independent mold) is capable of synthesizing methylmercury from the methionine biosynthetic pathway and mercuric salts (Landner 1971). Since many molds and photosynthetic bacteria are capable of synthesizing ethylene from methionine, it seems likely that this methylmercury is formed by the well-known reaction between ethylene and mercuric ion. This reaction was probably responsible for the chemical synthesis of contaminating methylmercury at Minamata (Kurland et al. 1960).

DEMETHYLATION OF METHYLMERCURY

The demethylation of methylmercury by sediments, microorganisms isolated from sediments, soil, and fecal organisms has been well established (Furukawa and Tonomura 1971, 1972a, 1972b; Summers and Silver 1972; Summers and Sugarman 1974; Schottel et al. 1974; Tezuka and Tonomura 1976). The enzyme system has been partially purified and was shown to consist of a hydrolase and a reductase. The hydrolase hydrolyzes the mercury-carbon bond to produce methane and mercuric ion. The reductase reduces the mercuric ion to mercury metal, which is volatilized from the aqueous culture medium into the atmosphere. Microbial resistance to methylmercury poisoning, as well as inorganic mercury poisoning, has been shown to develop through the transfer of extra-chromasomal factors (episomes) which carry the genetic information needed to produce both the hydrolase and reductase enzymes.

In purified enzyme systems, rate constants have been obtained for methylation (DeSimone et al. 1973) and demethylation (Tezuka and Tonomura 1976). Based on these initial studies, it appears that under ideal conditions demethylation is several orders of magnitude slower than methylation.

KINETICS OF METHYLMERCURY BIOSYNTHESIS AND METHYLMERCURY DIFFUSION INTO CELLS

The rate of methylmercury biosynthesis in selected ecosystems is determined by the microbial community

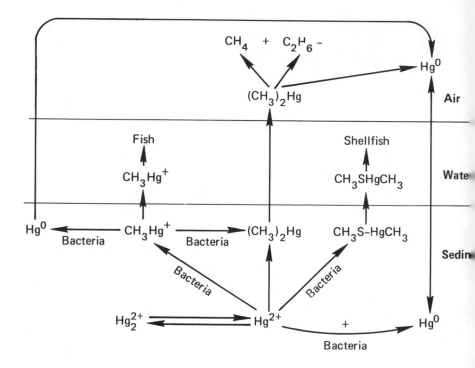

SOURCE: Modified from Wood (1974). Science 183:1049-1052. Copyright 1974 by the American Association for the Advancement of Science.

FIGURE 3.1 The mercury cycle demonstrating the bioaccumulation of mercury in fish and shellfish.

together with the available concentrations of soluble mercuric ion species and methyl-B_{12} compounds. The presence of demethylating microorganisms will allow a steady-state concentration of methylmercury to build up in any ecosystem, but the concentration would be lower than if they were less abundant. As with other elements on the earth's crust, mercury has a biological cycle (see Figure 3.1). Therefore, potential environmental problems with mercury develop only if the steady-state concentration of methylmercury increases in a particular environment (Wood 1976).

Evidence is now available to show that the bioaccumulation of methylmercury into the tissues of higher organisms (e.g., fish) is probably controlled by diffusion (S. Rakow and J.R. Lakowicz 1977, manuscript in preparation. Membrane permeability to methylmercuric chloride by fluorescense quenching. University of Minnesota, Freshwater Biological Institute). It has been determined that methylmercury chloride diffuses through cell membranes (40 angstroms) into cells in 20×10^{-9} s (Rakow and Lakowicz cited above). Once methylmercury diffuses through the cell membrane, it is rapidly bound by sulfhydryl groups (see Chapter 5, section entitled Protective Mechanism), thereby maintaining the concentration gradient across the membrane. This means that extremely low concentrations of methylmercury will bioaccumulate rapidly in ecosystems, thus explaining the bioaccumulation phenomenon observed for methylmercury in the environment, as well as the low steady state concentration found in water.

SUMMARY AND CONCLUSIONS

The biological synthesis of methylmercury is a natural phenomenon, and methylmercury is an integral part of the mercury cycle. The rate of synthesis of methylmercury is determined by the size of the natural microbial populations capable of synthesizing this neurotoxin. Microbial population sizes are determined by the extent of nutrient supplies to lakes, rivers, and coastal areas (i.e., the extent of eutrophication). Cleansing our waterways can certainly have a major effect on methylmercury synthesis in polluted areas, as could cutting back on the point sources for soluble inorganic mercury compounds.

Finally, once formed, methylmercury diffuses rapidly into the cells of higher organisms. The rates for bioaccumulation are so rapid that even low concentrations of methylmercury in water lead to elevated concentrations in fish.

CHAPTER 4

THE ECOLOGICAL EFFECTS OF MERCURY

Although the acute toxic effects of inorganic and organic mercury compounds on plants and animals have been determined, the effects from chronic low-level exposure have not been widely documented because these effects may not be apparent for a long time and the bioassay procedures currently available are insufficiently sensitive and selective to monitor them. Although low-level exposure to mercury has been associated with reduced growth and photosynthesis in phytoplankton and with teratogenic and mutagenic malformations in plants and animals, more precise tests and additional data are needed.

Most non-fish-eating terrestrial animals are not exposed to high levels of mercury and generally do not appear to be at risk except in regions of naturally occurring mercury deposits or in anthropogenically polluted areas. However, in most aquatic systems the organisms are constantly exposed to very low levels of methylmercury in water and food. If the watercourse receives anthropogenic mercury pollution, these low levels may increase and methylmercury may be concentrated to hazardous levels. Aquatic organisms absorb and magnify methylmercury at each trophic level of the food chain, thereby threatening not only the ecosystems but also human health, via the food chain. The results of one study of biological mercury magnification in the aquatic food chain are presented in Table 4.1.

An overview of the current information on the levels of mercury in terrestrial and aquatic ecosystems is presented here. Unless otherwise noted, the mercury concentrations are reported on a wet-weight basis. The levels in the atmosphere, soils, and water are discussed in Chapter 2 and levels in foodstuffs are discussed in Chapter 5.

44

TABLE 4.1   Biological Magnification of Mercury in the Aquatic Food Chain

	Number of Samples	Range of Values ($\mu g/g$)	Arithmetic Mean ($\mu g/g$)	More Numerous Organisms
Algae Eaters	39	0.01–0.18	0.05	Zooplankton; Snails; Mayfly nymphs
Zooplankton Eaters	9	0.01–0.07	0.04	Insect larvae; Minnows
Omnivores	9	0.14–1.16	0.45	Insect larvae and adults; Scuds
Detritus Eaters	12	0.13–0.89	0.54	Worms; Clams; Insect larvae
Predators	25	0.01–5.82	0.73	Insect larvae and adults; Frogs

SOURCE: Hamilton (1971).

MUSEUM SPECIMENS AS POLLUTION INDICATORS

Analyzing several species of preserved birds, Swedish researchers discovered that alkylmercury fungicides introduced to treat seed grain in the early 1940s had increased the mercury levels in the feathers of seed-eating birds. These results motivated other investigators to analyze additional preserved specimens. When four late Pleistocene mammals were tested, no mercury was detected in the bone marrow of a horse. Tissues from the foot of a mammoth, however, tested at 0.06 $\mu g/g$, and muscle tissue from a bison and moose showed 0.70 and 0.95 $\mu g/g$, respectively (Harris and Karcher 1972).

The Ontario Ministry of the Environment has analyzed two major sets of museum fish, one from Lake St. Clair and consisting of about 600 preserved fish covering the period 1853 to 1969, the other comprising 100 fish from Lake Simcoe covering the interval 1914 to 1968 (J.N. Bishop, personal communication, Laboratory Services Branch, Ontario Ministry of the Environment, Rexdale, Ontario, 1977). The former study indicated that fish captured after 1959 had significantly higher mercury levels than those from pre-1959 times. The Simcoe study indicated no change for the only species adequately represented (whitefish).

Isolated specimens of Pleistocene mammals without modern equivalents are only of historical interest. However, efforts have been made to compare the mercury burdens of preserved fish collected 25 to 100 yr ago for the purpose of determining whether levels of mercury are increasing over time in particular species. Most of these studies, however, are likely to be of more academic than practical importance because of insufficient data and possible contamination of the samples. The major source of error is in the preservation process. Comparisons of mercury levels in museum specimens and fresh or frozen fish must be considered unreliable until the changes that result from preservation are more fully understood. Mercury can be introduced as a trace impurity with the preservative or on metal tags or other identification materials placed in the container with the fish. Gibbs et al. (1974) found that the mercury content of eight commercial brands of formaldehyde ranged between undetectable and 9 $\mu g/l$ while metal identification tags contained between 30 and 50 μg Hg/g. Therefore, to provide a minimum degree of statistical confidence in the data, a number of preserved specimens of the same species, size, and age should be collected at each sampling location at selected time intervals; and the preservation and storage histories should be properly recorded. This might be done for future comparisons, but most current studies fall woefully short of these conditions and can only confirm that mercury does appear in samples of fish taken at earlier times.

Very few preserved marine fish have been analyzed, and arguments continue over whether any reliable conclusions can be drawn from the limited data (Barber et al. 1972, Miller et al. 1972, Saperstein 1973, Gibbs et al. 1974). More freshwater fish have been analyzed, and those caught recently average more mercury than preserved specimens, although this depends heavily on the degree of pollution, size of the fish, and the year and location where they were collected (Evans et al. 1972, Harris and Karcher 1972). The limited data presently available indicate that a small mercury burden may be normal and of natural origin, especially in marine fish. The levels in marine fish appear not to have changed to any measurable degree since the 1870s; the higher levels in freshwater fish are usually the result of additions from anthropogenic sources. However, care must be exercised in extrapolating the results of a very limited number of analyses of specimens with unknown preservation and storage histories. They lack enough statistical confidence to draw comparisons that indicate the degree and extent of mercury pollution in the present marine ecosystem.

ALGAE, PLANTS, AND INVERTEBRATES

Algae

Aquatic organisms derive most of their energy from algae at the base of the aquatic food chain. Very low levels of mercury can kill some species of algae and inhibit the growth of others, thereby disrupting the complex interdependency between phytoplankton and the populations that feed upon them. Furthermore, from this initial point of entry, the element is concentrated up the food chain.

Algae accumulate and concentrate mercury from the water primarily by surface absorption and to a lesser degree by adsorption (Hannerz 1968, Glooschenko 1969). The process of uptake is essentially passive. Freshwater algae and macrophytes were shown to accumulate radioactive mercury-203 in aquarium experiments by Makhonina and Gileva (1968).

Organomercurials retard the growth and viability of several species of marine algae more effectively than inorganic mercury (Boney et al. 1959, Boney and Corner 1959, Matida et al. 1971). As little as 0.1 μg/l of some alkylmercurial fungicides decreased the growth and photosynthesis of laboratory cultures of the marine diatom Nitzschia delicatissima as well as some freshwater phytoplankton. Panogen®, MEMMI®, and phenylmercuric acetate (PMA) were lethal to both marine and freshwater phytoplankton at 50 μg/l while less than 0.6 μg/l drastically limited their growth (Harriss et al. 1970). PMA inhibited the growth of three phytoplankton species at concentrations as low as 0.06 μg/l (Nuzzi 1972). Boney and Corner (1959) concluded that algae may be highly susceptible to alkylmercurials because of the high lipid content of their cell membranes while Matson et al. (1972) established that mercuric chloride and methylmercuric chloride inhibited the biosynthesis of lipids in selected species.

Young algae spores exposed to mercury may be retarded in growth as well as in the formation and maturation of reproductive structures, placing them at a competitive disadvantage with other species (Boney 1971). In addition, the response of phytoplankton to mercury toxicity has been related to other chemicals in the water. Increasing concentrations of other nutrient elements in the water can decrease the accumulation of mercuric chloride (Matida et al. 1971, Hannan and Patouillet 1972), and, conversely, extremely low levels may be lethal if other nutrients are insufficient (Harriss et al. 1970).

Plants

Aquatic Plants

Although aquatic vegetation contains much more mercury than the surrounding water, its total mercury level is relatively low. Plants in unpolluted waters usually measure between 0.03 and 0.08 μg/g total mercury while similar plants in polluted waters have concentrations up to 37 μg/g.

While most of the mercury in an aquatic system is associated with the sediments, higher plants form the largest biomass component in many watercourses, and they are capable of removing mercury from the water. The uptake of both inorganic and organic mercury is proportional to the length of exposure and the concentration (Fang 1973, Mortimer and Kudo 1975). Mortimer and Kudo found methylmercury to be more toxic than mercuric chloride for the aquatic plant Elodea densa. Fang demonstrated that Ceratophyllum demersum and Elodea canadensis metabolized PMA into mercuric mercury and small amounts of methylmercury and ethylmercury, and that the biological half-life of mercury residues ranged between 43 and 58 days.

Terrestrial Plants

All plants appear to concentrate traces of mercury,
but the amount depends on the plant species and
locality, and the chemical form of the mercury
available. The latter is most important. Rooted plants
absorb elemental mercury and alkylmercurials much more
easily than they do ionic inorganic mercury (Dolar et
al. 1971). Generally, total mercury concentrations in
most common edible plants and foods derived from plants
range from less than 1.0 to 300 ng/g, with the higher
levels due to natural deposits in the soil.
Regardless of the mercury concentration in the soil,
plants in well-aerated soils normally contain less than
0.1 to 0.7 μg/g on a fresh-weight basis. Plants have a
barrier to the uptake and circulation of inorganic
mercury salts and organically complexed mercurials
adsorbed on clay, as in aerated soils, but apparently
many plants have no barrier against the uptake of
gaseous mercury through the roots. In soils where
decaying sulfides release gaseous elemental mercury, the
vegetation has from 0.2 to 10 μg/g on a dry-weight
basis. In reducing soils, physical and chemical forces
firmly hold the mercury in insoluble sulfides or organic
complexes so that, typically, only from 0.01 to 0.04
μg/g is available to concentrate in the plants (Kothny
1973).

Invertebrates

Insects and Freshwater Invertebrates

Depending on the level of pollution, many invertebrate
organisms tolerate and magnify various mercurials to
some degree. In unpolluted Swedish waters, the level of
mercury in caddisflies (Trichoptera), stoneflies
(Plecoptera), alderflies (Neuroptera), and hogslater
(Asellus) ranged between 0.025 and 0.072 μg/g. In
waters polluted with phenylmercuric compounds, the
levels were approximately 100 times greater and ranged
from 1.9 to 17 μg/g (Johnels et al. 1967, 1968).
Although Hannerz (1968) found no direct correlation
between an invertebrate's mercury burden and its trophic
level, he did observe that predaceous insect larvae,
such as dragonflies and alderflies (Sialis), accumulated
more mercury than organisms which feed on decaying
plants or detritis. The extent to that organisms
concentrated mercury from water varied from <100 to
>12,000-fold, depending on such factors as the form of
the mercury, time of exposure, and nature of the food
consumed, as well as the feeding habits and metabolic

activity of the respective organisms. They accumulated more methyl, methoxyethyl, and phenylmercury than inorganic mercuric ion. In some cases, the insect's trophic level was less important than the quantity of mercury available for uptake (Hannerz 1968). Among larger invertebrates, the accumulation of mercury also depends on the compound as well as the species. Table 4.2 summarizes the concentrations among inhabitants of a pond ecosystem a month after it was treated with a single application of various mercury compounds.

Marine Invertebrates

Accumulation. In highly contaminated marine environments, invertebrates can concentrate large amounts of mercury. Mercury levels in crayfish in the Wabigoon River system in Ontario and in Minamata Bay ranged from 0.87 to 35.7 μg Hg/g compared to a range of 0.09 to 0.27 μg Hg/g for crayfish in unpolluted waters. Oysters (undisclosed species) at Minamata contained 5.61 μg Hg/g (Kitamura 1968) while oysters from Galveston and Chesapeake bays contained between 0.30 and 1.00 μg Hg/g (Kurland et al. 1960). In contrast, the average mercury content of rock oysters (Crassostrea commercialis) cultivated in relatively unpolluted industrialized and nonindustrial areas of Australia contained between 0.003 and 0.017 μg/g (Hussain and Bleiler 1973).

Toxicity. Much of the current information about marine invertebrates deals with acute mercury toxicity as it varies among species and also within a particular species according to its life stage. Larvae are generally most sensitive (Connor 1972).

Lethal doses of inorganic mercuric salts have been shown to relate directly to the species among invertebrates and range between 0.05 and 1800 mg/l (Jones 1940, Clark 1947, Barnes and Stanbury 1948, Pyefinch and Mott 1948, Hunter 1949, Wisely and Blick 1967, Knapik 1969). Corner and Sparrow (1957) found that brine shrimp and barnacle larvae were more sensitive to organomercurials than mercuric chloride and the toxicity levels reflected the rate of uptake for both species. Test animals absorbed fat-soluble organomercurial compounds into fatty tissue more rapidly than inorganic mercuric chloride. Among the various compounds containing radioactive mercury-203, n-amylmercuric chloride was 20 times more toxic to barnacles than mercuric chloride and 100 times more toxic to brine shrimp (Corner and Sparrow 1957, Corner and Rigler 1958).

Elimination. The patterns of elimination of radioactive mercury compounds were similar for fish,

TABLE 4.2   Concentration Factors After a Month Following a Single Dose
of Various Mercury Compounds into a Pond

	Mercuric Chloride	Methoxyethyl Hydroxide	Phenylmercuric Acetate	Methylmercuric Hydroxide
Water plants (six species)				
Submerged parts[a]	4–264	68–771	40–2350	34–3200
Emergent (three species)	3–77	4–53	8–90	8–25
Algae	252	920	1220	——
Moss	393	560	3900	5900
Invertebrates (four species)	247–560	510–1990	900–4200	3290–8470
Sediment	359	743	6800	6100

[a]Much of the variation is due to interspecific variation within the plants studies. For *Iris pseudacorus*, the values for the four compounds ranged from 4 to 98, while for *Lysimachia nummularia* the range was 264 to 3200.

SOURCE: Derived from Hannerz (1968).

freshwater crabs, other crayfish, and molluscs. The
biological half-lives for mercuric and phenylmercuric
nitrate injected into the foot muscle of a freshwater
mussel were 23 and 43 days, respectively. The
corresponding half-lives for methylmercuric nitrate
varied between 86 and 435 days, with the shortest half-
life in the youngest animal (Miettinen et al. 1969).
Although the quality of the water did not affect the
animal's rate of excretion, temperature did. The half-
life was estimated to be as long as 2 yr in summer and 3
yr in winter.

Crayfish appear to retain more mercury and eliminate
it more slowly when it is injected into muscle instead
of the mouth, apparently because there the mercury is
bound to protein in the tissues and central nervous
system. Only 4 to 20 percent is rapidly eliminated
according to Miettinen et al. (1968, 1969, 1972b). They
estimate that the half-life of mercury in the bodies of
the crayfish can range from 50 to 500 days depending on
the temperature of the water and route of
administration. Overall, an organism's retention of
mercury depends primarily on the form and means of
accumulation. These data are summarized toward the
bottom of Table 4.3.

FISH

Freshwater and marine organisms and their predators
normally contain more mercury than terrestrial animals.
In general, the uncontaminated background levels for the
flesh of animals, birds, and eggs are less than 0.02
μg/g. Levels of mercury in freshwater organisms and
fish taken from areas free of known mercury
contamination range from 0.02 to 0.2 μg/g. Levels in
top predatory fish such as bass (Micropterus dolomieui),
pike (Esox lucius), and walleye (Stizostedion vitreum)
range from 0.4 to 1.0 μg/g. However, mercury levels
vary widely depending on the age, weight, length,
species, metabolic rate, degree of pollution, locale,
and in some cases sex (Forrester et al. 1972, Olsson
1976, Bishop and Neary 1977). Moreover, statistical
evaluation is seldom adequate because the values are
usually based on small sample sizes, composite samples
from many different watercourses, or extrapolations from
the literature. Data on mercury levels in aquatic
organisms are difficult to compare primarily because
analytical uncertainties (Westoo 1975), poor
experimental designs, and biased interpretations of the
results have produced many ambiguities and
contradictions (see Appendix A).

The predominance of methylmercury in fish (80 to 90 percent as compared with inorganic mercury) has been demonstrated by numerous analyses (Bache et al. 1971, Kamps et al. 1972, Westoo 1973, Bishop and Neary 1976), although lower percentages of methylmercury have occasionally been reported for some fish (Bache et al. 1973, Freeman and Horne 1973). Body burdens are elevated substantially where the watercourses are contaminated with mercury from industrial and mining processes as well as organic and inorganic nutrients from anthropogenic sources.

Accumulation and Elimination

Fish accumulate mercury in excess of the 0.5 $\mu g/g$ FDA interim guideline from waters with low natural levels as well as from those that are anthropogenically contaminated. This accumulation is part of a dynamic process in which an organism's body burden strives to maintain equilibrium between intake and elimination. As discussed in Chapter 3, bioaccumulation occurs when the rate of uptake exceeds that of elimination, and this depends on a number of physical, chemical, and biological parameters that vary in each aquatic ecosystem (Burrows et al. 1974, Jernelov et al. 1975).

Uptake

The mercury accumulated in fish comes primarily through absorption from the water across the gills or through the food chain (Boetius 1960, Hannerz 1968, Hasselrot 1968, Amend et al. 1969, Backstrom 1969, Rucker and Amend 1969, Olson et al. 1973, Olson and Fromm 1973, Uthe et al. 1973, de Frietas et al, 1974, Hasselrot and Gothberg 1974), although some higher species may convert inorganic mercury into methylmercury (Westoo 1968, Imura et al. 1972). Some mercury is also taken up through the fish's mucus layer and/or skin (McKone et al. 1971, Burrows et al. 1974). The short chain alkylmercurials are absorbed much faster than alkoxyalkyl, aryl, or ionic inorganic mercurials (Hannerz 1968, Backstrom 1969).

Comprehensive data are not available on the feeding habits of fish. However, Jernelov (1972) and Jernelov and Lann (1971) estimated that fish at the higher trophic levels obtain half of their mercury burden through their food and the remainder from the water; fish at the lower trophic levels receive as much as 75 percent of their mercury burden from the water. On the other hand, Hannerz (1968) reported that animals

TABLE 4.3 Biological Half-life of the Slow and Fast Component of Excretion for Ionic and Protein-Bound Radioactive Mercurials

Aquatic Organism	Mercury Form	Method of Administration	Duration of Experiment (days)	Temp °C	Fast Excretion as Percent of Original Activity (days)*	Slow Excretion Biological 1/2 Life (days)*	Reference
Flounder (Pleuronectes flesus)	MMN	POP	100	13–19	33±4	780±120	a
Flounder (Pleuronectes flesus)	MMN	POI	100	13–19	12±4	700±50	a
Flounder (Pleuronectes flesus)	MMN	IMI	100	13–19	—	1200±400	a
Flounder (Pleuronectes flesus)	MMN	OBS	28	13–19	17±3	169±61	b
Flounder (Pleuronectes flesus)	MMN	OBS	77	13–19	23±2	430±115	b
Flounder (Pleuronectes flesus)	PMN	OBS	28	13–19	28±10	58±8	b
Flounder (Pleuronectes flesus)	PMN	OBS	77	3–19	40±7	164±16	b
Perch (Perca fluviatilis)	MMN	OBS	9–28	13–19	12±5	112±21	b
Perch (Perca fluviatilis)	MMN	OBS	77	3–19	8±8	470±37	b
Perch (Perca fluviatilis)	MMN	OBS	77	3–19	30	≥470	b
Perch (Perca fluviatilis)	PMN	OBS	34	3–19	32	157	b
Pike (Esox lucius)	MMN	POP	130	13–19	~10	750±50	a
Pike (Esox lucius)	MMN	POI	130	13–19	~5	640±120	a
Pike (Esox lucius)	MMN	IMI	130	13–19	—	780±80	a
Pike (Esox lucius)	MMN	OBS	20	16–19	10±5	140±40	b
Pike (Esox lucius)	MMN	OBS	84	3–19	11±1	490±30	b
Pike (Esox lucius)	PMN	OBS	21–53	11–18	35±41	64±4	b
Pike (Esox lucius)	PMN	OBS	84	3–18	30	190	b
Roach (Leuciscus rutilus)	MMN	OBS	8	15–17	12±4	25±9	b
Sea Bass (Serranus scriba)	MMN	OBS	60	17–23	4–6	267±27	c
Eel (Anguilla vulgaris)	MMN	POP	130	13–19	~17	910±40	a
Eel (Anguilla vulgaris)	MMN	POI	130	13–19	~17	1030±70	a
Eel (Anguilla vulgaris)	MMN	IMI	130	13–19	—	1030±80	a

Mollusc (Tapes decussatus)	MMN	IMI	99	17-23	20	481±40	c
Mollusc (Tapes decussatus)	MC	IMI	—	—	—	~100	d
Mollusc (Tapes decussatus)	MC	ASW	—	—		10 days	d
Mollusc (Tapes decussatus)	MC	RSF	—	—		5 days	d
Mollusc (Lymnaea stagnalis)	MMN	OBS	14-29	18±2	12	27±13	b
Mollusc (Mytilus galloprovincialis)	MMN	IMI	98	17-23	20	1000	c
Mussel (Pseudanodonta complenata)	MMN	IMI	275	13±6	7	435	b
Mussel (Pseudanodonta complenata)	MMN	IMI	63-106	17±3	10±3	129±22	b
Mussel (Pseudanodonta complenata)	MMN	IMI	54-106	17±3	9±5	86±10	b
Mussel (Pseudanodonta complenata)	PMN	IMI	6-40	8±2	3±3	43±7	b
Mussel (Pseudanodonta complenata)	MN	IMI	12-15	8±2	3±3	23±6	b
Crayfish (Astacus fluviatilis)	MMN	IMI	63	18±2	5	297	b
Crayfish (Astacus fluviatilis)	MMN	OBS	14-63	18±2	13±4	144±37	b
Crab (Carcinus maenas)	MMN	IMI	35	17-23	4-6	400±50	c
Seal (Pusa hispida)	MMN	POP	153	—	20	~500	e

*Error given as mean deviation from mean.

MMN—methylmercury nitrate
PMN—phenylmercuric nitrate
MC—mercuric chloride
MN—mercuric nitrate
POP. per os, proteinate-bound

POI—per os, free ionic form
IMI—into muscle, ionic form
ASW—absorption from sea water
RSF—resorption from food
OBS—orally administered into stomach of organism

References: a – Järvenpää et al. (1970).
b – Miettinen et al. (1969).
c – Miettinen et al. (1972b).
d – Unlü et al. (1972).
e - Keckes and Miettinen (1972).

bioaccumulate mercury in direct relation to such factors as metabolic rate and food habits rather than position on the trophic scale. Since a fish's metabolic rate is regulated by the temperature of the water, more mercury is concentrated in the summer than in the winter (Hannerz 1968, Hasselrot 1968, MacLeod and Pessah 1973, Uthe et al. 1973, Blaylock and Huckabee 1974, Reinert et al. 1974). Similarly, because of its higher metabolic rate, red muscle tissue takes up mercury faster than white muscle (Backstrom 1969). The metabolic rate of the fish and the mercury concentration in the aquatic ecosystem appear to be more important factors in bioaccumulation than age or exposure rate (Johnels et al. 1967, 1968; Olsson 1976).

For inorganic mercurials such as mercuric nitrate, concentration factors of between 1 and 50 have been measured for fish whereas for methyl and methoxyethyl mercury the organs and tissues of fish had values between 1000 and 2500 (Hannerz 1968). Estimates of experimental bioconcentration factors for methylmercury range from 3000 for pike (Johnels et al. 1967) to more than 8000 for rainbow trout (Salmo gairdneri) (Reinert et al. 1974). In addition, Underdal and Hastein (1971) reported biological magnification factors ranging from 7000 to 10,000 for rainbow trout downstream from a wood pulp factory.

The amount of mercury the fish are able to tolerate depends on the rate of uptake as well as the quantity and form administered. Miettinen et al. (1969, 1972a) showed that the levels of methylmercury in brain and nerve tissues tended to remain constant or even to increase without being supplemented, whereas phenylmercury remained in the digestive organs and was excreted much more rapidly.

Elimination

Large concentrations of mercury found in fish are caused not only by high rates of absorption but also by slow rates of elimination. A two-phase elimination process was observed in laboratory studies: a relatively fast phase when the fish's mucus layer sloughs off, and a slower second phase attributed to losses from muscle tissues (Massaro and Giblin 1972, Burrows et al. 1974). Dilution by growth of new flesh also affects the half-life of mercury in living organisms.

The rate of elimination depends largely on the specific compound and the metabolic rate of the fish. Matida et al. (1971) determined that rainbow trout eliminated methylmercuric chloride more slowly than ethylmercuric phosphate, phenylmercuric acetate, or mercuric chloride. When mercury intake ceases, warmer temperatures may speed the rate of elimination, while the growth of new tissue further decreases the mercury concentrations (MacLeod and Pessah 1973).

While the biological half-life of mercury can vary widely depending on its chemical form and the species of aquatic organism, in general, a half-life of at least 2 yr is normal for fish in the natural environment (Lockhart et al. 1972, Bishop and Neary 1974, Hasselrot and Gothberg 1974, Laarman et al. 1976). Table 4.3 presents the biological half-lives of mercury for various aquatic species, in terms of both the slow and fast phases of excretion.

Freshwater Fish

Analyses conducted by Stock and Cucuel (1934) and Raeder and Snekvik (1941) established that in unpolluted areas both marine and freshwater fish normally concentrate in the range of 0.04 to 0.15 μg/g mercury, concentrations being generally higher in predatory than bottom feeding species. After Swedish researchers began to analyze for methylmercury in the 1960s, they found elevated mercury levels in fish from contaminated fresh waters. Subsequently, levels in excess of 1 μg/g were reported in fish from Norway (Underdal 1969), Denmark (Dalgaard-Mikkelsen 1969), Canada (Wobeser et al. 1970; Bligh 1970, 1971; Bishop and Neary 1976; MOE 1977), Finland (Aho 1968), Italy (Ui and Kitamura 1971), and the United States (Bails 1972, Henderson et al. 1972, Henderson and Shanks 1973). Where waterways are highly contaminated, average mercury levels in adult predatory fish may approach 10 μg/g, and individual fish may exceed 24 μg/g wet weight (Bishop and Neary 1974).

In Canada the mercury content of fish varies widely. In 1975 Bishop and Neary (1976) determined the mercury levels in over 11,000 fish representing 19 species from 47 contaminated and uncontaminated lakes in northwestern Ontario. In all of the interlake comparisons, fish sampled covered a wide range of lengths and weights, so data were normalized to allow comparisons of specified lengths and species from one lake to another, and from one year to another for a given lake. In contaminated lakes the mean normalized mercury concentrations ranged from 0.40 to 1.21 μg/g for 25-in. pike, 0.54 to 1.32 μg/g for 20-in. walleye, and 0.04 to 0.28 μg/g for 20-in. whitefish (<u>Coregonus</u> <u>clupeaformis</u>). The ranges were even wider for anthropogenically contaminated lakes and waterways: from 2.87 to 5.25 μg/g for 25-in. pike, 1.97

to 6.90 $\mu g/g$ for 20-in. walleye, and 0.55 to 0.77 $\mu g/g$ for 20-in. whitefish.

Two of the most seriously polluted Canadian watercourses are the Wabigoon-English-Winnipeg river system in northwestern Ontario (Fimreite and Reynolds 1973, Annett et al. 1975) and the Lebel-sur-Quevillion region of northwestern Quebec (Barbeau et al. 1976). In the lakes and rivers polluted by a chloralkali plant near Lebel-sur-Quevillion, the mercury content of predatory fish such as pike and walleye has ranged between 0.08 and 4.88 $\mu g/g$ since 1971. The highest known levels of mercury in the Western Hemisphere were reported for fish taken from Clay Lake on the Wabigoon-English-Winnipeg river system. In 1970 the mean mercury levels were as follows: pike, 9.24 $\mu g/g$ (range 3.79 to 14.9); walleye, 12.1 $\mu g/g$ (range 1.2 to 24.0); burbot (Lota lota), 21.95 $\mu g/g$ (range 19.1 to 24.8); whitefish, 3.58 $\mu g/g$ (range 0.15 to 12.57); and white sucker (Catostomus commersoni), 3.83 $\mu g/g$ (range 1.68 to 7.97) (Bishop and Neary 1976). Only at Minamata and Niigata, Japan, were higher mercury levels reported in fish and shellfish from waterways that received direct discharges of methylmercury (Kitamura 1968, Tsubaki 1971).

On the Wabigoon River and Clay Lake the average mercury concentrations of some species of fish exceeded 10 $\mu g/g$ in 1970. The highest levels reported for walleye, pike, and burbot were 24.0, 27.8, and 24.8 $\mu g/g$, respectively (Fimreite and Reynolds 1973, Bishop and Neary 1976). For some lakes on the Wabigoon-English-Winnipeg river system no measurable decreases in the mean mercury levels have been observed for selected species of fish since 1970, while for selected species of fish from other lakes on the same system mean concentrations have decreased as much as 50 percent (Bishop and Neary 1976). More dramatic decreases were reported for fish taken from Lake St. Clair, the international waterway between Canada and the United States, after the major source of pollution, a chloralkali plant on the St. Clair River at Sarnia, Ontario, sharply reduced discharges in 1970 and totally eliminated them in 1975. While it is commonly believed that biomethylation of inorganic mercury was primarily responsible for the presence of methylmerucry in Lake St. Clair, the Ontario Water Resources Commission (Jernelov et al. 1972) established that a transalkylation of mercury with alkyl-lead compounds was an important source of both methyl- and ethylmercury to the system. Later, the Ontario Ministry of the Environment (MOE 1977) compared mercury concentrations in fish of similar size taken from Lake St. Clair between 1970 and 1976; the pattern of reduction closely approximates an exponential decline. In 1970 a 40-cm walleye had a mean mercury level of 2.1 μg/g whereas by 1976 a similar fish contained only 0.56 μg/g, a 73 percent reduction.

The Canadian researchers observed similar declines for all of the species tested from Lake St. Clair. American scientists have also observed these decreases but report a somewhat less dramatic decline rate. This discrepancy may not be significant because of the differences in the age, length, and weight of the fish considered and the statistical interpretation of the data. The average mercury concentration in yellow perch (Perca flavescens), rock bass (Ambloplites rupestris), and channel catfish (Ictalurus punctatus) dropped below the 0.5 μg/g FDA interim guideline by 1972; but some American scientists suggest that 5 yr or more will be needed for walleye 56 cm or larger to reach this level (W.A. Willford, personal communication, U.S. Fish and Wildlife Service, Great Lakes Fishery Laboratory, Ann Arbor, Michigan, 1977). Large adult walleye decreased their average mercury concentrations approximately 51 percent, from 2.83 μg/g in 1970 to 1.38 μg/g in 1972. From 1972 to 1975 the average mercury content decreased an additional 25 percent, but at a much slower rate, and this slowed decline continues.

Such rapid decreases may not occur in areas where contamination is removed less quickly. Lake St. Clair is a very shallow lake, and the approximately 8- to 9-day turnover is unusually rapid. A strong river current and the propellers on large numbers of lake freighters quickly translocate the fine sediment to Lake Erie. When Thomas et al. (1975) compared mercury concentrations in the sediments of Lake St. Clair in 1970 and 1974, they concluded that 64 percent had been transported downstream to Lake Erie. While the levels of mercury in the Lake Erie sediments have not decreased substantially since 1970, a significant decrease in the mercury levels in the fish has been observed (James N. Bishop, personal communication, Laboratory Service Branch, Ontario Ministry of the Environment, Rexdale, Ontario, Canada, 1977). Therefore, the rapid decline in mercury levels in fish is probably due to the combined effects of the elimination of the major source of mercury pollution, the rapid turnover time in the waterway, and the translocation of mercury contaminated sediments out of the lake.

Marine Fish

As with freshwater fish, mercury levels in marine fish vary markedly depending on the species, metabolic rates, age, weight, and geographical locations. Not all of the responsible factors have yet been explained, but average mercury levels are usually below 0.10 μg/g except for large carnivorous fish at the end of a long food chain, such as tuna and swordfish. Their mercury levels may exceed 1 μg/g (Doi and Ui 1975). However, where marine organisms were exposed to methylmercury from industrial effluent discharged directly into the ocean, fish and shellfish bioaccumulated up to 40 μg/g. These levels were reported at Minamata Bay, Japan, in the early 1950s (Kitamura 1968) and at Niigata, Japan, in 1965 (Tsubaki 1971).

Marine fishes generally have much lower total mercury levels than freshwater fish, and the ratio of inorganic mercury to methylmercury also varies more widely and may depend on the species. Almost all of the mercury contained in freshwater fish is in the form of methylmercury. The same is true of tuna and swordfish (Berglund et al. 1971, Doi and Ui 1975). However, in Pacific blue marlin (Makaira ampla) taken off the Hawaiian coast, only about 25 percent of the total mercury is in the form of methylmercury (Rivers et al. 1972).

In most cases, the mercury levels in marine fish seem to depend on where the fish were caught. In the United Kingdom, fish from the Thames estuary and eastern Irish Sea contained the highest levels, 0.45 to 0.5 μg/g and 0.55 to 0.64 μg/g, respectively. The mean concentration for fish caught within 25 miles of the coast was 0.29 μg/g but varied widely from area to area. Further offshore, the fish averaged 0.11 to 0.21 μg/g, and fish from the open sea ranged between 0.03 and 0.06 μg/g. Overall, the mean mercury concentration for fish and shellfish in the United Kingdom was estimated at 0.08 and 0.13 μg/g, respectively (U.K. Department of the Environment 1976). In 1973 cod, haddock, and plaice contained average mercury concentrations of between 0.04 and 0.13 μg/g. These three species accounted for 61 percent of the total fish caught in the United Kingdom.

The mercury levels in cod (Gadus morrhua) varied according to where they were taken. Those from the heavily contaminated waters of the western Baltic between Denmark and Sweden had values up to 1.29 μg/g. On the other hand, cod taken around Greenland had between 0.013 and 0.026 μg/g and North Sea cod ranged from 0.158 to 0.195 μg/g (Dalgaard-Mikkelsen 1969). In a similar study, Hall et al. (1976a) found that among 1227 Pacific halibut (Hippoglossus stenolepis) mercury concentrations in fish of the same size increased from the northern to the sourthern part of the species' ranges. Within each geographical area, the mercury concentration also increased with the size of the fish in a range from 0.15 to 0.45 μg/g. The element was

uniformly distributed throughout the edible muscle
tissues.

A study undertaken by the National Marine Fisheries
Service (Hall et al. 1976b) traced mercury levels in 205
species of finfish, molluscs, and crustaceans. These
species represented approximately 93 percent of the U.S.
commercial catch. The samples were taken from 198 sites
around the United States, including Alaska and Hawaii.
The mean levels of mercury exceeded 0.5 $\mu g/g$ in less
than 2 percent of the 159 species caught for human
consumption. Finfish represented 63.9 percent; most of
their muscles had a mean mercury level below 0.3 $\mu g/g$.
Only 31 species of finfish (1 percent of the total 63.9
percent) contained mean mercury levels that exceeded the
0.5 $\mu g/g$ FDA interim guideline.

Zook et al. (1976) surveyed 314 samples of 32 commonly
consumed fresh fish and shellfish for their total
mercury. The mean concentration was 0.12 $\mu g/g$. Only
three species--halibut, rockfish, and red snapper--
exceeded the 0.5 $\mu g/g$ level prescribed in the FDA
interim guideline. The highest levels were 0.52, 0.85,
and 0.55 $\mu g/g$, respectively. The 183 samples of
shellfish (oysters, clams, scallops, shrimp, and crabs)
averaged 0.05 $\mu g/g$, with only 10 samples exceeding 0.10
$\mu g/g$. The leg meat of a king crab was highest at 0.13
$\mu g/g$. Overall, the mean mercury level for the fish and
shellfish samples was 0.13 $\mu g/g$ with more than 96
percent testing below the FDA interim guideline of 0.5
$\mu g/g$. Similarly, among cod, clam, crab, flounder,
herring, lobster, and oysters taken off the Atlantic
coast in Canada, the mercury concentrations were 0.20
$\mu g/g$ or less (Bligh 1971).

As with the National Marine Fisheries Service study,
tuna and swordfish taken off the Atlantic coast of
Canada had higher mercury concentrations than other
fish. Concentrations ranged between 0.33 and 0.86 $\mu g/g$
and 0.82 and 1.00 $\mu g/g$, respectively (Bligh 1971).
Additional data are needed to determine whether these
species contain more mercury because of physiological
peculiarities or generally increased levels in the
marine ecosystem. Environmental pollution of their
habitat, as well as age, weight, metabolic rate, and
feeding habits, may all be contributing factors. An
age-dependent increase has been indicated by many
analyses. Furthermore, scombriformes species (spiney-
finned, as mackerels, tunas, bonitas, albacores, and
swordfish) have a high rate of metabolism and a high
level of food intake. Muscle tissue in tuna and
swordfish generally also contains about twice as much
dry matter as muscle tissue in other fish (Lofroth
1973). Among 88 yellowfin tuna (_Thunnus albacares_),
those that weighed less than 25 kg had mercury levels no

higher than 0.25 μg/g, those of 50 kg had mercury body burdens as high as 0.50 μg/g, while tuna that weighed between 60 and 100 kg had mercury levels as high as 1.0 μg/g. However, the larger fish also had a wide variation of mercury content (Doi and Ui 1975).

Various studies of the most common types of tuna have reported mercury levels that range widely and may depend on a number of factors. Among 911 samples of skipjack (Katsuwonus pelamis), yellowfin, and white tuna, mercury levels ranged from undetectable to 1.0 μg/g with most values between 0.20 and 0.30 μg/g. Most of the bluefin tuna (Thunnus thynnus)(285 samples) from the Bay of Biscayne had values near 0.5 μg/g in a range from 0.20 to 0.80 μg/g. The same species (136 samples) taken from the Mediterranean Sea ranged from 0.50 to 2.5 μg/g with most values close to 1.1 μg/g. Twenty samples of bigeye tuna (Thunnus obesus) of various origins had mercury levels ranging from 0.4 to 1.0 μg/g. Over 5200 samples of tuna (variety not specified) from Italy ranged from undetectable to 1.75 μg/g, with most mercury levels between 0.3 and 0.5 μg/g wet weight (U.K. Department of the Environment 1976).

BIRDS AND MAMMALS

Aquatic Birds

Determining the impact of contamination on aquatic birds is complicated because many of them live longer and migrate over a larger part of the hemisphere than their terrestrial counterparts. Exposure (including migratory patterns) and eating habits are the most significant factors in the concentration of mercury by aquatic birds. In seedeaters mercury levels can be reduced by eliminating mercury seed treatments. Plant eaters are not apt to concentrate much mercury from the little that is translocated. However, where fish concentrate extraordinarily high levels of methylmercury, so do birds that prey on them. The impact of mercury contamination has been assessed quite extensively; many investigators have reported on the content of liver and muscle tissues in aquatic bird species from various countries (Borg et al. 1969, Holt 1969, Koeman et al. 1969, Muto and Suzuki 1969, Environment Canada 1971, Fimreite et al. 1971, Johnson and Morris 1971, Koeman and van Genderen 1972, Vermeer and Armstrong 1972, Parslow 1973, Vermeer et al. 1973, Heath and Hill 1974, Annett et al. 1975).

Species of ducks that eat plants and seeds are least apt to concentrate mercury. The American widgeon (Anas americana) eats approximately 89 percent plant food

during its first 50 days of life, and the levels in five
of them averaged only 0.5 μg/g in the breast tissues
even along the highly contaminated Wabigoon-English-
Winnipeg river system. On the other hand, young
mallards (Anas platyrhynchos) have a 90 percent animal
diet, and 16 of them averaged 6.1 μg Hg/g in their
breast muscles. Blue-winged teal (Anas discors) eat
approximately equal parts of plant and animal food, and
their breast tissues ranged from 3.8 to 10.4 μg/g
(Vermeer et al. 1973).

In the same study along the Wabigoon-English-Winnipeg
river system, Vermeer et al. (1973) found that birds
that ate more fish had the highest mercury levels.
Among 21 common goldeneye (Bucephala clangula), the
average was 7.8 μg/g. Seventeen common mergansers
(Mergus merganser) averaged 6.8 μg/g, and seven hooded
mergansers (Lothodytes cucullatus) averaged 12.3 μg Hg/g
in their breast muscles. Crayfish in the stomachs of
hooded mergansers contained an average of 7.1 μg Hg/g in
1973. Five ducks were tested for methylmercury content
and had from 69 to 99 percent of the total mercury as
methylmercury.

Among puffins, guillemonts, cormorants, gannets,
gulls, ducks, and other fish-eating birds, investigators
measured mercury levels in liver ranging between 0.05
and 175 μg/g while muscle tissues ranged between 0.02
and 23 μg/g. In highly polluted areas fish eaters such
as herons, teal, mallard, and mergansers had levels in
muscle tissue between 0.51 and 23.0 μg/g (Dustman et al.
1972, Vermeer et al. 1973, Annett et al. 1975).

Migratory birds may detoxify if they spend part of the
year in uncontaminated areas. The osprey (Pandion
haliaetus), which migrates between Europe and the
Mediterranean or Africa, has a half-life of 2 to 3
months for methylmercury (Berg et al. 1966, Johnels et
al. 1968). Thus, feathers grown in a contaminated part
of central Sweden during June and July had 20 μg Hg/g
whereas those grown in October and November while the
birds wintered in the south had 5.0 to 6.3 μg/g. By
March the levels were down to between 1.8 and 2.3 μg/g.

On the other hand, along the heavily used Mississippi
Flyway, ducks may be exposed year-round if they summer
on a highly polluted waterway such as the Wabigoon-
English-Winnipeg river system and then migrate south to
winter on the polluted Galveston Bay. Ducks taken from
Lake St. Clair, North Dakota, and the Wabigoon-English-
Winnipeg river system have exceeded the FDA interim
mercury guideline of 0.5 μg/g; and in 1970, hunters were
advised not to eat the common goldeneye, blue-winged
teal, and mallard because of excess mercury in their
breast muscles (Environment Canada 1971).

Birds, such as the white-tailed eagle (<u>Haliaetus
albicilla</u>), that prey on fish but do not migrate are
unable to detoxify by changing food supplies.
Consequently, in 1965 some Swedish white-tailed eagles
had from 40 to 65 μg Hg/g in their feathers. Their eggs
seldom hatched after feather concentrations reached
between 5 and 11 μg/g. The greater incidence of
sterility corresponded with elevated mercury in the
birds' feathers, whereas those from uncontaminated areas
had feathers comparable to the nineteenth century museum
specimens.

The total mercury content of eggs from aquatic bird
species in various locales has been reported to range
between 0.03 and 15.8 μg/g (Eades 1966, Fimreite et al.
1971, Vermeer 1971, Wahlberg et al. 1971, Dustman et al.
1972, Faber and Hickey 1973, Greichus et al. 1973,
Holden 1973a, Blus et al. 1974). In contaminated areas
of Alberta, Saskatchewan, and Manitoba provinces, eggs
with from 0.9 to 22.7 μg/g hatched normally in 18 nests.
One fledgling's liver, however, contained 10 μg Hg/g
(Vermeer et al. 1973). In California, instead of
detoxifying their own systems by concentrating mercury
in their eggs, female gulls had mercury levels that
ranged 5.5 times higher than the 0.1 to 0.4 μg/g in
their eggs. Nonetheless, according to Wahlberg et al.
(1971), in isolated cases human populations may be at
risk when mercury contaminated wild duck eggs are
consumed.

Terrestrial Birds

Like fish, wild birds concentrate the highest levels
of mercury in the kidney and liver with less in the
muscle tissues. Although it is difficult to determine
the natural mercury levels for birds in uncontaminated
areas, Holden (1972) suggests numerous values below 0.1
μg/g in liver, kidney, and muscle tissues which indicate
that the natural levels for uncontaminated areas range
between 0.01 and 0.1 μg/g.

In the late 1950s Swedish ornithologists observed the
first mercury-related ecological problems. Many species
of birds, especially birds of prey, declined both in
numbers and breeding success. At the same time, mercury
levels increased in the feathers of several species of
seed-eating and predatory birds compared to preserved
specimens (Berg et al. 1966). Mercury levels in the
peregrine (<u>Falco</u> <u>peregrinus</u>) increased from a mean of
2.5 μg/g for the years 1834 to 1849 to more than 40 μg/g
for the years 1941 to 1965. Eagle owls (<u>Bubo</u> <u>bubo</u>) and
white-tailed eagles showed similar increases. For
goshawk (<u>Accipiter</u> <u>gentilis</u>) the average mercury content

was 2.2 µg/g prior to 1947, but between 1948 and 1965 the mean raised to 29 µg/g with some values in excess of 80 µg/g (Edelstam et al. 1969).

Seed-eating birds and their predators were widely contaminated by eating alkylmercury treated seed grain left uncovered during the spring and fall planting seasons (Johnels and Westermark 1969). Very high levels of mercury (up to 270 µg/g) were measured in the livers and kidneys of dead Swedish birds. Live birds had lower levels (1 to 53 µg/g)(Borg et al. 1966, 1969). After 1966 when the alkylmercurial seed dressings were replaced with alkoxyalkylmercury compounds, mercury concentrations declined significantly in seed-eating birds and their predators (Johnels and Westermark 1969). The mercury content of chicken eggs sold on the open market in Sweden also declined from 29 ng/g in 1964 to 9 ng/g in 1966 (Westoo 1969b), and similar decreases were observed in other foods. The mercury content of the liver and muscle tissues of terrestrial birds have been reported by many investigators (Borg et al. 1966, 1969, 1970; Eades 1966; Wanntorp et al. 1967; Holt 1969; Koeman et al. 1969; Fimreite et al. 1970; Anderson and Stewart 1971; Belisle et al. 1972; Huckabee et al. 1972; Martin 1972; Wiemeyer et al. 1972; Brock et al. 1973; Buhler et al. 1973; Griffith 1973; King and Lauckhardt 1973; Martin and Nickerson 1973; Smith 1973; Weigand 1973; Fimreite 1974; Kreitzer 1974).

In the United States and Canada, mercury-treated seed dressings also elevated the levels in seed-eating birds and their predators. In Canada the average mercury levels were 1.63, 1.88, and 1.25 µg/g for seed-eating songbirds, upland game birds, and rodents while the corresponding levels in similar specimens collected from an untreated area were significantly lower: 0.03, 0.35, and 0.18, respectively (Gurba 1970, Fimreite et al. 1970).

In 1970 both countries banned alkylmercurial seed dressings, and the levels decreased in game birds that do not feed on aquatic organisms. However, where phenylmercuric seed dressings continue to be applied in the United States, pheasants (Phasianus colchicus) and other wild birds can still accumulate up to 0.4 µg/g in the spring when the treated seeds are sown. The levels usually fall to less than 0.05 µg/g by summer.

Marine Mammals

At the top of the food chain, some larger marine mammals are showing high mercury concentrations in some organs (Helminen et al. 1968, Heppleston and French 1972, Sergeant and Armstrong 1973, Anas 1974). A number

of these animals have been tested along the California
coast and even in the Canadian Arctic where they were
presumed to be far removed from industrial pollution.
At Hudson's Bay, a beluga whale (Delphinapterus leucas)
had significant levels of total mercury in various
organs: muscle, 0.97 μg/g; liver, 8.87 μg/g; kidney,
2.44 μg/g; and heart, 1.35 μg/g (Bligh 1971). Levels of
mercury in the livers of six pilot whales (Globicephala
scammoni) stranded off the California coast ranged from
8.5 to 23.9 μg/g (Anonymous 1971). Male and female
harbor porpoises (Phocoena phocoena) taken from the Bay
of Fundy also had significantly different mercury
levels. Among 41 muscle and 20 liver samples, the total
mercury for male muscle tissues averaged 0.75 μg/g in a
range from 0.21 to 1.92 μg/g. Female muscle tissue
averaged 1.02 μg/g in a range from 0.26 to 2.58 μg/g.
The liver concentrations ranged from 0.89 to 18.30 μg/g
and 0.55 to 91.30 μg/g for males and females,
respectively. Essentially all of the mercury associated
with the muscle tissues was methylated whereas in the
livers, methylmercury represented from 7.4 to 41 percent
of the mercury present (Gaskin et al. 1972).

California sea lions (Zalophus californianus
californianus) have elevated mercury, DDT, and PCB
levels; and premature births have increased in their
breeding rookeries since 1968. Whether the cause is one
or all of these has not been determined, but females
have mercury residues ranging from 38 to 64 μg/g in
their livers (Delong et al. 1973).

Seals taken from widely distant areas have shown
mercury concentrations that exceed background levels.
Adult harp seals (Pagophilus groenlandicus) averaged
0.34 and 3.68 μg/g in the muscle tissue and liver,
respectively (Sergeant and Armstrong 1973). Harbour
seals (Phoca vitulina) from the Bay of Fundy and Gulf of
Maine had average total mercury levels of 0.054 μg/g
(range, 0.027 to 0.106), 0.59 μg/g (range, 0.16 to
1.54), 8.75 μg/g (range, 0.52 to 50.9), and 0.27 μg/g
(range, 0.05 to 0.76) in their blubber, muscle, liver,
and cerebrum, respectively (Gaskin et al. 1973). Along
the Netherlands coast, harbor seals exposed to high
levels of mercury from industrial pollution had mean
mercury levels of 340 μg/g, 20 to 50 times more than the
values reported in other studies. These seals showed
definite pathological signs despite a strong correlation
between the mercury and selenium levels in their livers
(see Chapter 5) (Koeman et al. 1973).

In Scotland liver mercury levels in gray seals
(Halichoerus grypus) normally range from 10 to 50 μg/g,
but up to 720 μg/g were found in areas where the levels
in fish gave no indication of mercury contamination.
The highest values appeared mainly in very old seals

(Holden 1973a, 1973b). In Canada gray seals had up to
2.35 µg/g in the muscle and 387 µg/g in the liver
(Sergeant and Armstrong 1973). In British waters the
brains of both common seals and gray seals concentrated
0.3 to 0.7 µg Hg/g when they were from 12 to 18 months
old whereas the livers of older seals concentrated 4.9
to 113.0 µg/g. Seals from East Anglia and West Scotland
accumulated more mercury than those from the Outer
Hebrides, Shetland, and Farne islands (Heppleston and
French 1972).

In the Canadian Northwest Territory, ringed seals
(Phoca hispida) showed mean mercury levels of 27 µg/g in
liver and 0.53 µg/g in muscle. Bearded seals
(Erignathus barbatus) had 143 µg/g in their livers and
0.53 µg/g in muscle. The mercury content was positively
correlated with age and body weight, but methylmercury
was only a small fraction of the total, 5.6 and 0.38
percent, respectively (Heppleston and French 1972,
Sergeant and Armstrong 1973, Anas 1974, Smith and
Armstrong 1975).

One member of a relic population of ringed seals
(Phoca hispida saimensis) in Lake Saimaa, Finland, was
captured alive because of impaired coordination that
resembled the symptoms of mercury poisoning. The
animals had been feeding on fish containing 0.2 µg Hg/g,
and it had concentrated 210 µg/g in the liver and 197
µg/g in the flesh. Two other seals had 74 and 130 µg/g
in their livers (Henriksson et al. 1969). A 9-month-old
female ringed seal was fed radioactive (mercury-203)
methylmercury in herring to determine retention and
excretion rates. Some mercury was excreted as the body
levels increased within the first 3 weeks, but 40
percent redistributed in the body fat and was retained
for a biological half-life of 500 days, approximately
the same excretion rate as for other sea animals
(Tillander et al. 1972).

Terrestrial Mammals

The mercury burdens in terrestrial mammals usually are
directly related to their diets and are low compared
with marine mammals. Herbivores normally have the
lowest mercury levels while carnivores that prey on
aquatic organisms have the highest body burdens. Borg
et al. (1969) reported that herbivorous animals such as
the roe deer (Capreolus capreolus) and hares (Lepus
timidus, L. europaeus) had less than 1 µg/g in mixed
liver and kidney, while the mercury levels in
carnivorous animals such as marten (Martes martes),
polecat (Mustela putorius), and fox (Vulpes vulpes)
exceeded 30 µg/l in mixed liver and kidney. Johnels and

Westermark (1969) reported that herbivores such as the cow, horse, moose, and deer concentrate only from 0.007 to 0.075 µg Hg/g in their tissues. However, in Alberta, where organomercurial seed dressings were used, the livers of ground squirrel (Spermophilus richardsonii) averaged 1.05 µg/g. In Saskatchewan liver mercury concentrations in the same species averaged only 0.10 µg/g (Fimreite et al. 1970).

In 1972, 48 fur bearing animals from the Bell-Nottaway basin in northwestern Quebec had low mercury residues in muscle tissue. For example, beaver (Castor canadensis) ranged from 0.01 to 0.05 µg/g; otter (Lutra canadensis), 1.88 to 2.10 µg/g; wolf (Canis lupus), 0.38 µg/g; rabbit, 0.03 µg/g in muscle and 0.44 µg/g in kidney; muskrat (Ondatra zibethicus), 0.05 µg/g in muscle and 0.34 µg/g in kidney; and marten, 0.12 to 0.33 µg/g for muscle and 0.56 to 1.02 µg/g in kidneys (Environment Canada 1972). In other parts of Canada the following average total mercury concentrations in muscle and liver tissue were reported: moose (Alces alces), 0.04 µg/g (range <0.01 to 0.17); mule deer (Odocoileus hemionus), 0.13 µg/g (range 0.06 to 0.18); caribou (Rangifer tarandus), 0.017 µg/g in muscle and 0.20 µg/g in liver; wolf, 0.051 µg/g in muscle and 0.24 µg/g in liver; and Arctic fox (Alopex lagopus), 0.31 µg/g in muscle and 0.76 µg/g in liver (Smith and Armstrong 1975). In addition, muskrat taken from the Lake Erie forest district had liver mercury levels that ranged between 0.040 and 0.251 µg/g (Jervis et al. 1970). All of the mercury in 10 specimens of polar bear (Ursus maritimus) muscle tissue was in the form of methylmercury and averaged about 0.13 µg/g (range 0.01 to 0.66 µg/g)(Desai-Greenway and Price 1976).

CHAPTER 5

ENVIRONMENTAL EXPOSURE AND UPTAKE OF MERCURY BY HUMANS

The toxicity of methylmercury to humans is well known and is discussed in detail in Chapter 6. The uptake of this form of mercury presents a potentially serious hazard to health. The significance of environmental sources is increased by the fact that mercury methylation has not been shown to occur in vivo in humans. This chapter discusses ways in which people can be exposed to mercury compounds in the environment and, where the data permit, to methylmercury especially.

AIR AND DRINKING WATER

There is no indication that mercury compounds in the concentrations and forms found in either the atmosphere or drinking water supplies contribute significantly to the methylmercury burden in human beings. Most data indicate that the mercury levels in water are, with rare exceptions, less than 1 μg/l and almost always lower except in regions with anomalously high concentrations of mercury in the rocks and soils or near sources of anthropogenic pollution. For example, in 1971, EPA's Division of Water Hygiene analyzed 698 samples of raw and finished waters collected from 273 drinking water supplies throughout the United States. Only 11 samples exceeded 1 μg/l in a range from 1.0 to 4.9 μg/l, and only 1 sample exceeded 5 μg/l (Hammerstrom et al. 1972). A more recent survey of finished waters established that only 13, or 2.5 percent, of 512 mercury analyses exceeded the proposed 1975 federal drinking water standard of 2 μg/l (U.S. EPA 1973, 1975a). The World Health Organization (1971) has recommended an upper limit of 1 μg/l for the total mercury content of water for human consumption. Thus, assuming a daily intake of 2 liters of water and other beverages, a daily maximum of 4 μg can be attributed to this source. If the actual observed mercury concentration in most water supplies is considered (<0.1 μg/l), this source contributes less than 0.2 µg/person/day and is inconsequential when compared with the intake from foods.

FOOD SUPPLIES

General Food Surveys

The available data indicate that almost all the methylmercury in the human diet comes from fish, other seafood, and possibly red meat. However, other kinds of food contribute to the total human mercury burden. Minute amounts of mercury occur naturally in nearly all foods and beverages, and also as a contaminant from the use of mercury compounds as fungicides and in industry.

The amounts of methylmercury in edible plants and plant products are generally extremely low; exceptions may occur if the plants are grown on contaminated soil or from seed stock treated with mercury. Meat and dairy products may contain low levels of total mercury, which can include a proportion of methylmercury compounds presumably derived from residues in foods that contain fishmeal and, in the past, from mercury-treated cereal grains.

Mercury levels in foods have been surveyed in several countries. In most cases the analytical data report only the total mercury content. For example, Stock (Stock and Cucuel 1934, Stock 1938) reported that traces of mercury occur in nearly all foods. Several later food surveys are summarized in Table 5.1. In general, the following average mercury concentrations were reported for uncontaminated foods: dairy products, 0.002 to 0.020 µg/g; meat, fish, and poultry, 0.010 to 0.20 µg/g; grain and cereal, 0.020 to 0.050 µg/g; potatoes, 0.006 to 0.020 µg/g; legume vegetables, 0.002 to 0.010 µg/g; and beverages, 0.002 to 0.006 µg/g.

In the United Kingdom when over 6400 samples of food were analyzed (U.K. Ministry of Agriculture, Fisheries, and Food 1971, 1973; U.K. Department of the Environment 1976), the mean level of mercury in cereals, most fresh meats, fruits and preserves, green and root vegetables was less than 0.005 µg/g on a fresh-weight basis. The natural mercury content of herbage is about 0.001 to 0.010 µg/g. In addition, animals that eat food supplements prepared from soybeans, grains, fish, milk, meat, and bone may add mercury as follows: vegetable matter <0.01 to 0.07 µg/g, fish <0.01 to 1.8 µg/g (mean value <0.3 µg/g), milk 0.01 µg/g, and 0.01 to 4 µg/g from meat and bone. Since the mean values were 0.01 to 0.04 µg/g for vegetable feeds, 0.06 to 0.23 µg/g for fish feeds, and 0.37 µg/g for meat and bone meal, it seems reasonable to conclude that natural herbage and vegetable feeds are an insignificant source of mercury in animals. Fish meal, depending on the amount consumed, is more important because the mercury content is largely in the methyl form. Meat and bone meal supplements may also be important, but unlike the fish meal only a negligible portion of the mercury is present as methylmercury compounds (U.K. Department of the Environment 1976).

A 1970 survey of nearly 1400 Canadian foods, excluding fish, showed the mercury residues to be less than 0.060 μg/g for bread, flour, grains, and eggs, and less than 0.040 μg/g in meats and vegetables (Somers 1971). In Michigan the results of a limited survey showed that all foods except fish contain detectable traces of mercury in the range <0.01 to 0.03 μg/g (Gomez 1972). A series of market-basket studies reviewed by Berglund et al. (1971) established the average daily intake in the Scandinavian countries between 1965 and 1971. With occasional exceptions, i.e., 0.18 μg/g mercury in hog liver, the mean mercury level of foodstuffs was 0.03 μg/g or less.

In a survey of 10 food commodities by the U.S. Food and Drug Administration (FDA) in 1970, 1971, and 1972, shrimp had the highest mercury content at 0.043 μg/g and a median value of 0.014 μg/g (Simpson et al. 1974). Nonfat dry milk was next highest at 0.027 μg/g and a median of 0.010 μg/g. All other commodities (flour, sugar, potatoes, raw ground beef, chicken breast, beef liver, eggs, and fluid whole milk) had mean mercury concentrations of less than 0.003 μg/g. In the total diet fractions, only meat, fish, and poultry contained mercury as high as 0.041 μg/g (range from 0.004 to 0.041 μg/g). All other fractions (dairy products, grain and cereal products, potatoes, leafy vegetables, legume vegetables, root vegetables, garden fruits and fruits) contained average mercury levels of less than 0.002 μg/g on a dry-weight basis.

Of all food commodities analyzed for mercury, only fish appear to present a potential hazard to human beings. The 0.5 μg/g FDA interim guideline is usually only exceeded in the larger species (large tuna, halibut, and swordfish) of the commercially important marine varieties and in freshwater fish obtained from mercury polluted watercourses. Mercury levels in fish are reviewed in detail in Chapter 4 and more briefly in the next section. In all other foods, mercury either is not detectable or is present at levels approximately 2 orders of magnitude lower than the FDA interim guideline for fish.

TABLE 5.1 The Mercury Content of Selected Foods

Type of Food	Country	Mean (ng/g)	Range (ng/g)	Reference
DAIRY	United States	–	<1-3	Simpson et al. (1974)
PRODUCTS	United States	–	1-2	Kirkpatrick and Coffin (1974)
	Canada	–	2-20	Corneliussen (1969)
Milk	United States	–	5-20	Gomez (1972)
	United States	–	3-7	Stock and Cucuel (1934)
	United States	8	–	Goldwater (1964)
	United States	1	1-9	Tanner et al. (1972)
	Germany	–	6-4	Stock and Cucuel (1934)
	United Kingdom	10	ND-20	U.K. Ministry of Agriculture, Fisheries and Food (1971)
Powder	Canada	–	21-180	Jervis et al. (1970)
Nonfat-dry	United States	10	4-27	Tanner et al. (1972)
Cheese	United States	20	15-30	Gomez (1972)
	United States	80	–	Goldwater (1971)
	United Kingdom	170	ND-500	U.K. Ministry of Agriculture, Fisheries and Food (1971)
	Canada	70	–	Jervis et al. (1970)
Butter	United States	140	–	Goldwater (1971)
MEAT, FISH,	United States	–	4-41	Simpson et al. (1974)
AND POULTRY	United States	–	10-50	Corneliussen (1969)
	Canada	–	28-51	Kirkpatrick and Coffin (1974)
Meat, General	United States	–	0.8-44	Gibbs et al. (1941)
	United States	–	1.0-150	Goldwater (1964)
	United States	3	2-7	Tanner et al. (1972)
	United States	–	ND-20	Gomez (1972)
	United Kingdom	–	ND-90	U.K. Ministry of Agriculture, Fisheries and Food (1971)
	Canada	<40	–	Somers (1971)
	Germany	–	1-67	Stock and Cucuel (1934)
	Worldwide	–	3-30	Smart (1968)
Liver, Pork	United States	30	20-40	Gomez (1972)
	Canada	–	16-170	Jervis et al. (1970)
	Sweden	–	3-49	Westöö (1969a)
Liver, Beef	United States	3	<2-8	Tanner et al. (1972)
	United States	10	10-15	Gomez (1972)
	United Kingdom	40	ND-90	U.K. Ministry of Agriculture, Fisheries and Food (1971)
	Canada	–	14-199	Jervis et al. (1970)
Kidney	Germany	67	–	Stock and Cucuel (1934)
Shrimp	United States	14	5-43	Tanner et al. (1972)
Chicken	United States	20	15-30	Gomez (1972)
	Worldwide	–	5-21	Smart (1968)
Breasts	United States	3	1-7	Tanner et al. (1972)
	Canada	–	25-61	Jervis et al. (1970)
Eggs	United States	ND	–	Gibbs et al. (1941)
	United States	–	10-62	Goldwater (1971)
	United States	<2	2-5	Tanner et al. (1972)

TABLE 5.1 (continued)

Type of Food	Country	Mean (ng/g)	Range (ng/g)	Reference
	United States	30	20-40	Gomez (1972)
	United Kingdom	–	ND	U.K. Ministry of Agriculture, Fisheries and Food (1971)
	Canada	–	20-29	Jervis et al. (1970)
	Canada	<60	–	Somers (1971)
	Germany	2	–	Stock and Cucuel (1934)
	West Germany	–	5	Smart (1968)
	Sweden	–	4-21	Westöö (1969a)
	Sweden	29	15-43	Smart (1968)
	Norway	–	15-20	Smart (1968)
	Denmark	–	4	Smart (1968)
	Italy	–	5-6	Smart (1968)
	Austria	–	6-13	Smart (1968)
	Belgium	–	6-8	Smart (1968)
	Holland	–	5-7	Smart (1968)
GRAINS AND	United States	–	<2-12	Simpson et al. (1974)
CEREAL	United States	–	20-50	Corneliussen (1969)
PRODUCTS	Canada	–	2-4	Kirkpatrick and Coffin (1974)
Barley	United States	25	20-40	Gomez (1972)
Millet	United States	30	–	Bache et al. (1973)
Oats	United States	10	5-10	Gomez (1972)
Rice	United States	87	–	Gerdes et al. (1974)
	United States	10	5-10	Gomez (1972)
	United Kingdom (imported)	–	5-15	Smart (1968)
	Japan	–	80-190	Fujita (1966)
Wheat	United States	20	10-30	Gomez (1972)
	Sweden	–	8-12	Smart (1968)
	Worldwide	–	20-85	Jervis et al. (1970)
Grain	United States	–	2-6	Gibbs et al. (1941)
	United States	–	2-25	Goldwater (1964)
	Canada	<60	–	Somers (1971)
	Germany	–	20-36	Stock and Cucuel (1934)
Flour	United States	3	3-6	Tanner et al. (1972)
	United States	103	92-118	Gerdes et al. (1974)
	United Kingdom	20	ND-60	U.K. Ministry of Agriculture, Fisheries and Food (1971)
	Canada	<60	–	Somers (1971)
	Germany	–	25-35	Stock and Cucuel (1934)
Bread	United States	10	5-10	Gomez (1972)
	United Kingdom	20	ND-80	U.K. Ministry of Agriculture, Fisheries and Food (1971)
	Canada	96	5-10	Jervis et al. (1970)
	Canada	<60	–	Somers (1971)
	Germany	5	–	Stock and Cucuel (1934)

TABLE 5.1 (continued)

Type of Food	Country	Mean (ng/g)	Range (ng/g)	Reference
VEGETABLES	United States	ND	–	Gibbs et al. (1941)
	United States	–	ND-60	Goldwater (1964)
	United States	–	2-20	Goldwater (1971)
	United States	–	1-123	Gerdes et al. (1974)
	United States	–	10-35	Gomez (1972)
	United Kingdom	–	ND-100	U.K. Ministry of Agriculture, Fisheries and Food (1971)
	Canada	<40	–	Somers (1971)
	Canada	–	22-280	Jervis et al. (1970)
	Germany	–	2-44	Stock and Cucuel (1934)
Leafy	United States	–	<1-9	Simpson et al. (1974)
	Canada	1	–	Kirkpatrick and Coffin (1974)
	Worldwide	–	12-44	Smart (1968)
Root	United States	–	<1-2	Simpson et al. (1974)
	Canada	1	–	Kirkpatrick and Coffin (1974)
Legume	United States	–	<1-2	Simpson et al. (1974)
	United States	–	2-10	Corneliussen (1969)
	Canada	1	–	Kirkpatrick and Coffin (1974)
Potatoes	United States	–	<1-13	Simpson et al. (1974)
	United States	–	6-20	Corneliussen (1969)
	United States	3	1-15	Tanner et al. (1972)
	United States	6	–	Bache et al. (1973)
	United States	10	ND-15	Gomez (1972)
	United Kingdom	–	10-79	Lee and Roughan (1970)
	United Kingdom	20	ND-70	U.K. Ministry of Agriculture, Fisheries and Food (1971)
	Canada	–	200	Jervis et al. (1970)
	Canada	1	–	Kirkpatrick and Coffin (1974)
	Germany	1	–	Stock and Cucuel (1934)
	Worldwide	–	3-6	Smart (1968)
Tomatoes	United States	18	–	Bache et al. (1973)
	United States	10	ND-10	Gomez (1972)
	United Kingdom	10	ND-50	U.K. Ministry of Agriculture, Fisheries and Food (1971)
	New Zealand	–	12-144	Smart (1968)
FRUIT	United States	4	4-30	Goldwater (1964)
	United States	–	<1-3	Simpson et al. (1974)
	United States	–	ND-282	Gerdes et al. (1974)
	United States	10	ND-10	Gomez (1972)
	United Kingdom	–	ND-180	U.K. Department of Agriculture, Fisheries and Food (1971)
	Canada	–	24-270	Jervis et al. (1970)
	Canada	1	–	Kirkpatrick and Coffin (1974)
	Germany	–	4-10	Stock and Cucuel (1934)
	Worldwide	–	3-50	Smart (1968)

TABLE 5.1 (continued)

Type of Food	Country	Mean (ng/g)	Range (ng/g)	Reference
MISCELLANEOUS				
Sugar	United States	51	–	Gerdes et al. (1974)
	United States	<3	<3-10	Tanner et al (1972)
	United States	10	5-10	Gomez (1972)
Beer	United States	4	–	Goldwater (1964)
	Germany	–	0.07-1.4	Stock and Cucuel (1934)
Fat				
Vegetable	Germany	–	60-115	Stock and Cucuel (1934)
Animal	United Kingdom	140	20-260	U.K. Ministry of Agriculture, Fisheries and Food (1971)
	Germany	–	70-280	Stock and Cucuel (1934)

ND = Not Detected.

Mercury Levels in Fish

In 1970, 1971, and 1972 the FDA surveyed (1) the mercury content of a wide variety of fish samples from selected freshwater regions, commercial fish from wholesale distributors, and swordfish and canned tuna fish; (2) 10 commodities representing a high proportion of total food consumption; and (3) 12 total diet fractions collected in the FDA continuing market-basket study to determine pesticide residues in the basic 2-week diet of a 19-year-old male (Simpson et al. 1974). The survey found that some freshwater fish species contained elevated mercury levels traceable to known sources of pollution. In 1970, over 20 percent of the fish sampled from Lake Erie exceeded 0.5 μg/g; over 50 percent of the fish from the Ontario side of Lake St. Clair were also above this level. The fish that usually contained the highest mercury levels were walleye, sheepshead, smallmouth bass, white bass, catfish, perch, and carp. Of domestic samples collected in 1971-1972 from all parts of the country, 16 of 40 samples (40 percent) had mercury levels in excess of 0.5 μg/g. Similar results were observed for imported freshwater fish, with 7 of 16 samples tested (or about 45 percent) containing more than 0.5 μg/g mercury (Simpson et al. 1974).

A number of commercial fish--halibut, bonita, mackerel, cod, and red snapper--contained slightly elevated levels but averaged below the 0.5 μg/g FDA interim guideline. For example, 500 samples of halibut showed the overall average concentration of mercury to be approximately 0.25 with about 13 samples above the 0.5 μg/g guideline. Of more than 1300 samples of 19 different species of commercial fish and seafood collected in 1971, only six species included one or more fish with mercury levels greater than the 0.5 μg/g guideline. Bonita and red snapper averaged about 0.30 μg/g wet weight. They were the only species with a significant number of samples that averaged more than 0.20 μg/g. In addition, one study of fish protein concentrates and fish meals manufactured from North American fish showed a mercury range (dry weight) from 0.3 to 0.9 μg/g (Beasley 1971).

While the mercury level of most marine and freshwater fish is less than 0.20 μg/g, some large marine fish such as tuna, swordfish, marlin, halibut, and shark usually range from 0.20 to 1.50 μg/g and can be as high as 5.0 μg/g (wet weight). Swordfish showed the highest incidence of mercury. Over 95 percent of the 853 swordfish samples analyzed in 1971 exceeded the 0.5 μg/g FDA interim guideline, with more than 50 percent exceeding 1.0 μg/g mercury (wet weight). Swordfish from

the western Atlantic (210 samples) had mean mercury
levels of 1.15 μg/g wet weight (ranging from 0.050 to
4.9 μg/g) while 49 swordfish taken near Italy had levels
ranging from 0.65 to 1.75 μg/g with most values near
1.10 μg/g wet weight (U.K. Department of the Environment
1976). When 3000 samples of canned tuna were tested,
the average total mercury content was approximately 0.25
μg/g with only about 4 percent of the samples exceeding
the 0.5 μ/g FDA interim guideline.

The overall mean value of 0.02 μg/g was confirmed for
a variety of canned fish such as salmon, herrings,
pilchards, sardines, and mackerel, while the mean
mercury levels in tuna ranged between 0.07 and 0.44
μg/g. For canned shellfish such as shrimp, prawns,
crab, and lobster, the mean mercury levels ranged
between 0.01 and 0.29 μg/g for samples taken from
various waters (U.K. Department of the Environment
1976).

Human Uptake of Mercury in Fish

Fish and shellfish are the only regular source of
methylmercury in the human diet of practical importance
today. Elevated mercury and methylmercury levels in
human beings have been linked with fish consumption
(Berglund et al. 1971, Suzuki et al. 1971, Yamaguchi et
al. 1971). Where selected subpopulations and
individuals eat large amounts of fish from highly
contaminated local waterways, either freshwater or
marine, over time a health hazard could develop (see
Chapter 6). Even without anthropogenic contamination,
fish and shellfish contain more mercury than other
foods. Marine fish and shellfish taken from unpolluted
waters typically contain 0.01 to 0.3 μg/g mercury on a
wet-weight basis. Individual fish taken from moderately
polluted waters have mercury levels in the range of 0.5
to 3 μg/g. In heavily contaminated areas such as
Minamata Bay and Niigata, Japan, the afflicted victims
ate fish and shellfish that contained up to 35 μg/g,
mostly in the form of methylmercury (Kitamura 1968). In
North America, fish containing mercury concentrations in
excess of 24 μg/g (wet weight) have been taken from the
heavily contaminated Wabigoon-English river system in
northwestern Ontario (Bishop and Neary 1976). This
situation poses a serious health hazard to tourists and
especially to native Americans who continue to eat fish
from these waters (Takeuchi et al. 1977).

In some areas of the United States, mercury
contamination of fish has become severe enough to
require health warnings or closure of sport and
commercial fisheries. By November 1970, 11 states

(Alabama, Georgia, Louisiana, Michigan, Mississippi, New York, Ohio, South Carolina, Tennessee, Texas, and West Virginia) had issued health warnings and closed both or either their sport or commercial fisheries because of excessive mercury contamination. An additional 6 states (New Hampshire, New Mexico, Pennsylvania, Vermont, Virginia, and Wisconsin) issued warnings to the general public about the health hazards from eating mercury-contaminated fish (Harlan 1971). Since then, 6 states (Alabama, Georgia, Louisiana, Mississippi, Ohio, and West Virginia) have reopened their fisheries while 6 others (Michigan, New York, Tennessee, Texas, Vermont, and Virginia) currently have some waters closed to fishing because of mercury pollution. The most recent fisheries closure occurred in Virginia on June 6, 1977. In addition, 12 states (California, Idaho, Illinois, Kentucky, Massachusetts, Minnesota, Mississippi, New Mexico, North Carolina, Oregon, South Carolina, and South Dakota) have current health warnings with respect to eating mercury contaminated fish from certain state waterways. A more complete survey of the current status of individual state sport and commercial fisheries is given in Appendix C.

In the Hackensack Meadows area of New Jersey a major mercury pollution problem is being investigated by the New Jersey Department of Environmental Protection. Soil, sediments, and water contain high levels of total mercury: percents in soil, 10's of $\mu g/g$ in sediments, and ng/ml in the water. However, the levels in fish are well below the FDA interim guideline, and there is no immediate threat to human health. However, this could change if the mercury becomes mobilized and moves into the food chain, so it is a major concern (Dr. Peter W. Preuss, personal communication, Special Assistant to the Commissioner, N.J. Department of Environmental Protection, August 10, 1977).

Consumption Patterns and Estimates

A widely quoted National Marine Fisheries Service study of fish consumption patterns (Nash 1971) measured only how much fish was purchased in 1969-1970. The fish were grouped in general categories such as tuna, salmon, or halibut and such fish specialty items as tuna pie, clam chowder, and TV dinners. The study assessed the per capita fish consumption only by socioeconomic class with no consideration of extremely heavy fish eaters, but it did provide evidence that overall fish consumption had increased 13 percent by 1973. Canned tuna increased 30 percent, canned salmon decreased 33 percent, and fish and frozen fish increased 45 percent.

Among other things, prices and changing dietary habits, ethnic or religious customs influence fish consumption patterns. Jews eat more fish than other religious groups, for example, while Catholics eat more than Protestants. Blacks eat more fish than whites (Nash 1971).

The 1973-1974 Seafood Consumption Study (National Purchase Diary Panel, Inc. 1975) estimated sport catches and servings eaten away from home, as well as commercial purchases. The study assumed an average dinner portion to be 6.0 oz (170 g), less than an average Weight Watchers® portion (8 oz, 227 g, for women and 10 oz, 284 g, for men); and a lunch and snack portion to be 3.5 oz (99 g) again less than the Weight Watchers® 4-oz (113-g) portion. Thus, consumption rates were figured at less than normal portions and minimum mercury levels.

Among the 7999 families comprising 26,848 individuals studied in 1973-1974, 93 percent consumed seafood, with tuna the most common fish consumed by 61.5 percent. The average total consumption was 18.6 oz (527 g) per month with a maximum of 247 oz (7002 g). The maximum tuna consumption was 202 oz (5727 g). The tuna samples showed a mean mercury concentration of 0.225 μg/g with a standard deviation of 0.12. Pooling cans produced a standard deviation of 0.42 μg/g for a single can of 6.5 oz (184 g). A consumption rate of 200 oz (5670 g) per month would produce a median intake of approximately 42.5 μg/day at the rates analyzed in December 1970 (National Purchase Diary Panel, Inc. 1975). Few individuals, however, consumed this much. While 38.5 percent of the populace ate no tuna, the average for the remainder was 6.5 oz (184 g) per month; only 2.8 percent ate more than 18.0 oz (510 g), and 0.5 percent more than 38 oz (1077 g) per month. Most other fish eaten by high tuna consumers contained approximately 0.1 μg/g mercury, according to the limited available data (Marsh et al. 1975).

The amount of mercury ingested varies from day to day and from one individual to another because the mercury concentrations vary in different types of foods and in samples of the same type of food, and because people's eating habits also vary widely. It is very unlikely that all foods in a normally mixed diet will consistently have high levels of mercury over long periods of time. In Sweden, Westoo (1965) established that the mean mercury content of 12 fish-free daily diets in Stockholm was 10 μg mercury per day with a range of 4 to 19 μg. More recently, Dencker and Schutz (1971) established the mean mercury intakes for 17 individuals to be 3.6 μg (range 1.0 to 9.3 μg) while the corresponding mean level for 58 diets containing fish was 8.7 μg/person/day, with a range of 1.7 to 30.6 μg mercury.

Tolerance Limits and Guidelines

The total dietary intake of mercury in foodstuffs varies widely from one country to another, primarily because of differences in fish consumption rather than other foodstuffs. Table 5.2 presents per capita seafood consumption estimates and limits of total mercury concentrations in food for various countries. Guidelines or tolerance limits for acceptable levels of mercury in fish and other seafood are usually based, at least in part, on the average rates of daily fish consumption.

The average seafood intake for Japanese citizens is 109 g/day, by far the highest daily intake of the countries listed; and the maximum mercury level declared acceptable in fish is 0.4 μg/g. The problem of mercury in Japanese food is exacerbated by the combination of an exceptionally high level of fish consumption where some fish are very highly contaminated in combination with the low average body weight (50 kg) of the average Japanese citizen.

In Europe, tolerance limits range from 1 μg/g for fish in Sweden to 0.03 μg/g for foods in general in the Benelux countries (Smart and Hill 1968). The United Kingdom has no statutory limits specifically for mercury residues in foods, although the average daily consumption of fish there varies from 12 to 34 g per person (U.K. Department of the Environment 1976) and contributes from 2 to 5 μg of mercury per person. The total mercury ingested by an average individual in the United Kingdom is about 5 to 10 μg/day, based on 1.5 kg of food eaten daily (35 to 70 μg/week or 0.07 to 0.14 μg/kg of body weight per day for an adult). About half of this is in the methylated form. These figures are about 4 to 8 times lower for total mercury and 6 to 12 times lower for methylmercury than the World Health Organization's provisional tolerable weekly intake (U.K. Department of the Environment 1976).

These values from the United Kingdom agree quite well with the daily intake of less than 14 μg reported by Abbott and Tatton (1970) for their total diet studies, as well as the daily intakes estimated by Stock and Cucuel (1934) of about 5.0 μg/person; by Gibbs et al. (1941) of about 20 μg/person; and by Monier-Williams (1950) of between 5 and 20 μg/person. The same range of values appears to be valid for the great majority of individuals in the United States. Peyton et al. (1975) estimated that for a standard diet the low and high daily intakes range between 5.4 and 14.6 μg total mercury of which 2.5 to 7.1 μg is methylmercury.

While food, and primarily fish, is the major source of mercury to humans, the average intake of methylmercury from this source is well below the WHO provisional tolerable limit. Other sources, such as skin preparations, dental fillings, and inhalation also contribute to the total mercury ingestion. The per capita total sorbed dose has been estimated to range from 17 to 41 μg mercury/person/day (Peyton et al. 1975).

First, a Swedish expert group (Berglund et al. 1971) and then the WHO recommended maximum intake levels to protect humans from risk through long-term ingestion of methylmercury. The Swedish authorities established 20 ng/g as an acceptable level of methylmercury in whole human blood with 30 μg as the acceptable daily intake (ADI). The World Health Organization (1972) accepted the findings of the Swedish authorities and set a provisional tolerable weekly intake of 300 μg of total mercury of which not more than 200 μg should be methylmercury.

Even though the recommended weekly intake of methylmercury for a 70-kg adult is essentially the same under both directives, the Swedish ADI and the WHO provisional tolerable intake should be clearly distinguished. The WHO defines an acceptable daily intake as the intake of a chemical (expressed in mg of the chemical per kg of body weight) which, during an entire lifetime, appears to be without appreciable risk on the basis of all the known facts at the time. The phrase "without appreciable risk" is taken to mean the practical certainty that injury will not result even after a lifetime of exposure. The ADI's set by WHO and FAO of the UN are intended to allocate the acceptable amounts of an additive where it will serve specific, necessary purposes in accordance with good manufacturing practice. Since such concepts are not applicable for metallic contaminants such as mercury or methylmercury in food, the Joint FAO/WHO Expert Committee on Food Additives (WHO 1972) adopted a new approach by allocating a provisional tolerable weekly intake of food contaminants. The term "tolerable" signifies permissibility rather than acceptability since the intake of mercury or methylmercury is unavoidably associated with the consumption of otherwise wholesome and nutritious food, and the term "provisional" indicates that this evaluation is tentative. (See Appendix B: A Brief Review of the FAO/WHO Deliberations on Setting Acceptable Tolerances for Mercury Residues in Foods, 1963-1976.)

TABLE 5.2 National Per Capita Seafood Consumption Estimates and Limits for Total Mercury Concentrations in Foods for Various Countries[a]

Country	Seafood Consumption (g/person/day)[b]	Food Regulated	Total Mercury Limit (μg/g)	Reference
Australia				
South Australia	17	Not specified	0.1	FAO/WHO (1968)
Victoria	17	Not specified	0.1	FAO/WHO (1968)
Western Australia	17	Not specified	0.01	FAO/WHO (1968)
Benelux	37	Not specified	0.03	FAO/WHO (1968)
Brazil	9	Not specified	0.05	FAO/WHO (1968)
Canada[c]	16	Fish	0.5	Morrison (1971)
Denmark	53	Not specified	0.05[c]	FAO/WHO (1968)
West Germany	30	Not specified	Zero[d]	FAO/WHO (1968)
		Fish (dogfish, tuna and swordfish)	0.5	
Japan[c]	109[f]	Fish	0.4[e]	Takeuchi and Eto (1975)
New Zealand	19	Not specified	Zero	FAO/WHO (1968)
		Fruits and vegetables	0.05	FAO/WHO (1968)

Sweden	56	Not specified	0.05[c]	FAO/WHO (1968)
		Fish	1.0	Berglund et al. (1971)
United Kingdom	26	All foods	None[g]	Portmann (1977)[g]
United States[c]	17	Fruits and vegetables	Zero	FAO/WHO (1968)
		Fish	0.5	Kolbye (1970)
		Water	0.002	U.S. EPA (1975b)
WHO/FAO		All foods	h	WHO (1972)

[a] Unless noted, the residues are measured as total mercury.

[b] FAO Food Balance Sheets, 1946-1966; Rome, 1971.

[c] Guideline only (maximum acceptable levels).

[d] Where derived from pesticide treatment.

[e] No more than 0.3 μg/g as methylmercury.

[f] From Takeuchi and Eto (1975).

[g] The United Kingdom Ministry of Agriculture, Fisheries and Food recommends that local enforcement officers and Public Analysts "exercise vigilance in ensuring that food containing levels of mercury unacceptable in their country of origin do not find their way on to the United Kingdom market" (J. E. Portmann, Ministry of Agriculture, Fisheries and Food, Fisheries and Food Laboratory, Burnham-on-Crouch, Essex, England, personal communication, 1977). The recommendation is aimed primarily at fish and shellfish and is designed to preclude dumping of condemned fish or shellfish by other countries.

[h] A "provisional tolerable intake" of 0.3 mg/person/week of which no more than 0.2 mg/person/week can be in the methylmercury form.

THE EFFECTS OF SELENIUM ON METHYLMERCURY TOXICITY

Nishigaki et al. (1974) analyzed 279 samples of 24 species of marine fish and found that the fish with higher levels of methylmercury generally also contained even higher levels of selenium (0.5 to 1.0 μg/g). Methylmercury in tuna, swordfish, and other large ocean fish appears to be less toxic than methylmercury ingested under other circumstances. This hypothesis has been proven experimentally in animals and attributed to the antagonistic effects of selenium, which is usually present in levels equal to or exceeding the mercury content of the marine organism.

The protective effect of dietary selenium against mercury toxicity was originally shown by Parizek and co-workers (Parizek and Ostadalova 1967, Parizek et al. 1969). It was subsequently confirmed by several other investigators (Ganther et al. 1972, 1973; Levander and Argett 1969; Stillings et al. 1972; Iwata et al. 1973; El-Begearmi et al. 1973, 1975; Groth et al. 1973; Potter and Matrone 1973, Welsh et al. 1973; Froseth et al. 1974; Ganther and Sunde 1974; Hill 1974; Johnson and Pond 1974; Stoewsand et al. 1974, 1977; Welsh and Soares 1974; Ohi et al. 1975; Ueda et al. 1975a, 1975b; Sell and Horani 1976).

Observed Protective Effects

The most consistent beneficial influence of selenium has been a reduction of the lethal and neurotoxic effects of methylmercury compounds. This was noted when the two were administered simultaneously in the diets of rats and Japanese quail. In addition, the depression of growth by methylmercury was partially alleviated. Ganther et al. (1972) found that Japanese quail (Coturnix coturnix japonica) fed 20 μg/g mercury in a corn-soya diet were intoxicated at 4 weeks and had a 52 percent mortality after 4 to 6 weeks. However, those fed the same concentration of mercury in a diet containing 17 percent tuna were free of symptoms of intoxication for a longer time, and only 7 percent died. Tuna contain enough selenium (normally up to 2 μg/g) to reduce the toxic level of methylmercury. Evidence that selenium protected against methylmercury intoxication in quail has also been reported by Stoewsand et al. (1974). Ohi et al. (1976) reported that naturally occurring selenium in tuna increased the growth rate of test animals intoxicated with methylmercury, but was only about half as effective in preventing the neurological symptoms of intoxication as was selenite selenium added to their diet. In studies of chronic dietary exposure

to methylmercury and selenite, control rats died with
symptoms of neurotoxicity and concentrations of
methylmercury in the brain well below those in rats
protected by selenite (Ueda et al. 1975a, 1975b). Given
parenterally, selenite decreased the toxicity of dietary
methylmercury and temporarily increased the
methylmercury concentration in the rat brain (Iwata et
al. 1973).

Protective Mechanism

The reason for the protective action of selenium is
unclear. Sumino et al. (1977) present evidence that
selenite releases methylmercury from its linkage with
proteins and thereby influences its tissue distribution.
In a person whose methylmercury body burden is high
enough to cause clinical symptoms, the presence of
selenium appears to immobilize the methylmercury
compound, but does not appear to speed its elimination
from the body (Stillings et al. 1974).
The natural biological sink for methylmercury is in
its interaction with sulfhydryl groups. In fact,
approximately 95 percent of the methylmercury bound to
fish protein is part of the methylmercury-cysteinyl
coordination complex. This complex can be broken up
with an excess of thiols or strong acids. Recently,
Japanese workers (Sugiura et al. 1976) have shown that
the selenohydryl group binds methylmercury 100 times
more tightly than the sulfhydryl group.
The important role of selenium as an essential trace
element in biological systems is also now well
documented (Stadtman 1974). Selenium forms the active
site of the enzyme glutathione peroxidase (Stadtman
1974) and readily replaces sulfur in the sulfur-
containing amino acids. In view of the fact that the
selenohydryl group binds methylmercury 100 times more
tightly than the sulfhydryl group, it is clear that
organisms with a diet supplemented by selenium or with
high natural levels achieve an added degree of
protection against methylmercury poisoning.
Some livers and brains of apparently healthy sea
mammals have high concentrations of mercury and
selenium. Koeman et al. (1973) found a strong 1:1
correlation between the concentrations of mercury and
selenium in the livers of seals and dolphins. Martin et
al. (1976) have shown a similar correlation among the
mercury, selenium, and bromine levels of female
California sea lions. In healthy animals the ratios
among all three elements are similar, but those animals
with an imbalance have a strong tendency to produce pups
prematurely. Since Tamura et al. (1975) established

that the selenium-to-mercury molar ratio in tuna and
bonito muscle tissue was 5.8:1, it appears that the 1:1
ratio is established internally whenever sufficient
selenium is present in the mammal's diet. These
findings are in very good agreement with the results of
experimental work on the binding of mercury and selenium
to the blood plasma proteins of animals given mercuric
chloride and selenite (Burk et al. 1974) and the
correlation between mercury and selenium reported for
human subjects exposed to inorganic mercury (Kosta et
al. 1975). Recently, Nishigaki and Harada (1975)
investigated the preserved umbilical cords of
individuals affected by Minamata disease and showed that
the mercury content was much greater than the selenium
content. For people not affected by the disease, the
mercury:selenium ratio was 1:1. Moreover, Birke et al.
(1972) reported on individuals with elevated blood
mercury levels who remained free of neurological
symptoms. This suggests that agents such as selenium
may modify the toxic effects of methylmercury and may
account for some of the variations in individual dose
response. However, the protective effect from selenium
has not been documented in humans. For this reason and
because the protective mechanism is unclear, it would be
premature to modify current guidelines for acceptable
levels of mercury in food on the basis of selenium
content.

Recent studies of human populations that consume large
quantities of tuna have revealed no definitive sign of
mercury poisoning, although some individuals had
elevated mercury levels in blood and hair. Whereas
marine fish such as tuna tend to accumulate both mercury
and selenium at approximately the same level in the
oceans, fish from polluted fresh waters concentrate much
more mercury because of the greater supply in relation
to selenium. Further studies should be initiated to
establish the role of selenium in the freshwater
ecosystem.

Other Protective Agents

While the current evidence suggests that selenium is
the main protective factor in tuna, other dietary
factors may also modify the expression of methylmercury
toxicity and must be taken into account by toxicologists
and regulatory agencies that set tolerance standards.
Vitamin E has been shown to have a similar protective
effect against methylmercury poisoning (Welsh 1974,
Harada et al. 1975, Welsh and Soares 1976, El-Begearmi
et al. 1976). However, a much higher concentration is
required to provide the same level of protection as with

selenium. The mechanism by which Vitamin E protects against methylmercury toxicity is not understood at the present time either. However, both glutathione peroxidase and Vitamin E remove activated oxygen species and may prevent radical attacks on methylmercury, which would be expected to give toxic free radicals as products (Ganther [In press]). Moreover, by adding selenium the protective effects of Vitamin E against methylmercury poisoning in quail can be enhanced. Vitamin E was most effective in diets containing 0 to 0.1 μg/g selenium. At these levels, supplemental Vitamin E significantly decreased the clinical symptoms of toxicity and improved the survival rate of the birds (Welsh and Soares 1976). In another series of experiments, El-Begearmi and co-workers (1974) showed that a combination of sodium arsenite and sodium selenite is more effective than either one separately. Other compounds actively being investigated for mutual antagonism toward methylmercury include: cystine (Stillings et al. 1974), methionine (El-Begearmi et al. 1974), Vitamin E (Welsh and Soares 1976), ascorbic acid (C.L. Farakas, personal communication, University of Waterloo, Waterloo, Ontario, Canada, 1977), and selenocysteamine (Sugiura et al. 1976).

CHAPTER 6

EFFECTS ON HUMAN HEALTH[1]

Methylmercury compounds have no known normal metabolic
function; their presence in the tissues of living
organisms, including man, represents contamination from
environmental sources. Ultimately, limits for the
release of any potentially toxic substance, including
mercury, into the general biosphere must be assessed in
terms of risk both to human populations that might be
exposed to it, and to the ecology. A useful cost-
benefit analysis of the use of mercury compounds and
their release into the environment would require
accurate and specific delineation of risk. Risk must be
defined in terms of the toxic effects of acute, as well
as chronic or lifetime, human exposure, including all
life-cycle stages. In addition to evaluating toxic
effects on the central nervous system, the potential
risks of adverse genetic, reproductive, teratogenic, and
carcinogenic effects of these nonessential compounds
must be determined before definitive human and
environmental tolerance levels can be estimated with
confidence.

DEFINITIONS

The critical concentration of a toxic substance in a
cell or organ is that level at which an adverse or
critical effect can be detected by clinical observation,
functional tests, or morphological or biochemical
techniques. Critical organ concentration is defined as
the mean concentration in an organ at the time any of
its cells reaches critical concentration. The critical
organ is the organ that first attains critical organ
concentration (Task Group on Metal Accumulation 1973).
Existing evidence strongly indicates that the critical
organ system in methylmercury (MeHg$^+$) poisoning of man
is the central nervous system (Berglund et al. 1971,
Bakir et al. 1973, Task Group on Metal Accumulation
1973, WHO 1976a, Tsubaki and Irukayama 1977). Symptoms
and signs observed in human populations after exposure

to toxic doses of methylmercury are dominated by
neurological disturbance. These effects may appear
weeks to months after acute exposure to toxic doses. A
progression of symptoms and signs after short- and long-
term exposure includes: paraesthesia (numbness and
tingling of the lips, mouth, hands, and feet),
dysarthria, ataxia, concentric constriction of the
visual fields, blurred vision, blindness, deafness, and
ultimately death.

Relationships between dose and degree of an effect
have not been quantitatively established for
methylmercury because of difficulties in quantitating
clinical effects. It has, however, been possible to
relate dose and appearance of various effects of
methylmercury. Doses have been estimated from ingested
amounts as well as index media such as blood or hair
concentrations. An indication of the primitiveness of
current ability to measure effect is that the critical
effect has to be defined as broadly as paraesthesia, a
symptom response reported by the exposed individual. No
other effect has yet been recognized at lower doses.

An effect is a defined biological reaction that may be
quantifiable and graded (degree of effect), or quantal
(all or none). Response is the frequency of an effect
in a population resulting from exposure to a specific
dose, i.e., an enumeration of reactors versus
nonreactors. Effect may be thought of as a change, due
to a dose, within an individual; response, the number of
different individuals of a total population showing a
specific effect (Task Group on Metal Accumulation 1973).
Thus, it follows that there is a population dose-
response relationship for each observable effect.

TOXIC EFFECTS OF METHYLMERCURY IN ADULT POPULATIONS

Hazard Estimates

Studies of human populations in Japan and, more
recently, in Iraq after large-scale, high-dose exposure
have documented the toxicity of methylmercury compounds
for human beings (Berglund et al. 1971; Bakir et al.
1973; WHO 1976a, 1976b; Tsubaki and Irukayama 1977).
Observations on these and other exposed populations have
been used to estimate the hazard of methylmercury
exposure for human populations.

Japanese Exposures: Contaminated Fish

The two major Japanese outbreaks of methylmercury poisoning in Minamata Bay and in Niigata were caused by industrial release of methylmercury and other mercury compounds into Minamata Bay and into the Agano River, followed by accumulation of methylmercury in edible fish and shellfish. The median total mercury level in fish caught in Minamata Bay at the time of the epidemic has been estimated as 11 μg/g fresh weight (Berglund et al. 1971). More recent reports of follow-up studies on these exposed populations have revealed many cases that were not originally diagnosed. By 1974, more than 700 cases of methylmercury poisoning had been identified in Minamata and more than 500 in Niigata (Tsubaki 1971, WHO 1976b, Tsubaki and Irukayama 1977).

The detailed risk-evaluation for long-term human exposure to methylmercury compounds performed in 1971 by a Swedish expert group (Berglund et al. 1971) was based on these Japanese exposures and used two independent approaches: a metabolic method and an epidemiological method. The metabolic evaluative approach was based on the lowest concentrations of mercury observed in the brains of fatal cases, absorption and distribution data, and assumption of a biological elimination half-time of 70 days. A critical daily intake was estimated to be 750 μg Hg as methylmercury for an average 70-kg person. Assuming daily ingestion, this is the level at which observable clinical effects would be expected to occur in sensitive adult individuals. The epidemiological assessment was based on the lowest blood concentrations in Japanese individuals observed to have clinical symptoms and signs, and on data relating levels of mercury in nonpoisoned individuals to the estimated daily intake of methylmercury consumed in fish. A critical daily intake was calculated to be 300 μg Hg as methylmercury for an average 70-kg person. A recent reassessment of hair and blood data from Minamata and Niigata discusses some of the difficulties of using these observations to estimate the minimal clinical effect level of methylmercury (Marsh et al. 1975).

In order to safeguard the health of the population, the lower of the two estimates of critical daily intake (i.e., 300 μg of methylmercury) was reduced by a safety factor of 10 to arrive at an acceptable daily intake (ADI). The Swedish expert group (Berglund et al. 1971) concluded that this safety factor was adequate to protect even the most sensitive members of a population such as children and pregnant women, that is, that clinical symptoms of mercury poisoning would not occur until a level of 200 ng/g was reached in the blood-- equivalent to 60 μg/g in the hair and a body burden of

20 to 30 mg for a 70-kg person. They suggested that 30 μg Hg as methylmercury be considered an acceptable daily intake for an average 70-kg person and 20 ng/g as an acceptable level of mercury as methylmercury in whole human blood.

Since the completion of the Swedish evaluation, significant new information is becoming available from additional studies of exposed human populations (Bakir et al. 1973, Clarkson and Marsh 1976, WHO 1976a). Ongoing studies of exposed Iraqi and Canadian native populations will provide additional important data for estimating human health hazard during the next few years.

Iraqi Exposure: Contaminated Bread

During the winter of 1971-1972, the largest outbreak of methylmercury poisoning ever recorded occurred in Iraq as a result of consumption of homemade bread prepared from seed wheat treated with a methylmercury fungicide. More than 6000 poisoned children and adults were admitted to hospitals throughout Iraq with nearly 500 reported hospital deaths (Bakir et al. 1973, WHO 1976a). It is probable that more people were poisoned and died without contacting the hospital system.

In the exposed Iraqi population, it has been possible for the first time to determine separate dose-response curves relating population frequencies of paraesthesia, ataxia, dysarthria, deafness, or death to body burdens of methylmercury at the time of appearance of the effect, estimated from quantity of contaminated bread ingested. Abrupt change in the slope of the observed frequencies of a specific effect with increasing body burdens has allowed an extrapolated estimate of the minimal mean body burden at the time the specific effect (the detection limit) could first be detected above mean background frequencies of the observed effect (Bakir et al. 1973). The limit of detection of this type of dose-response relationship is determined by the background frequency of the effect being measured in unexposed individuals in the population being examined. The most sensitive current index of human methylmercury toxicity is paraesthesia, which appears above background frequency in the population samples in Iraq at a mean body burden of 25 mg or 40 mg mercury as methylmercury (50-kg individual), depending on the factor used to convert intake to estimated body burden (Bakir et al. 1973; see Figures 6.1 and 6.2). Estimates of mean body burdens for appearance of other signs (ataxia, dysarthria, death) and symptoms (deafness) of methylmercury poisoning were similarly extrapolated

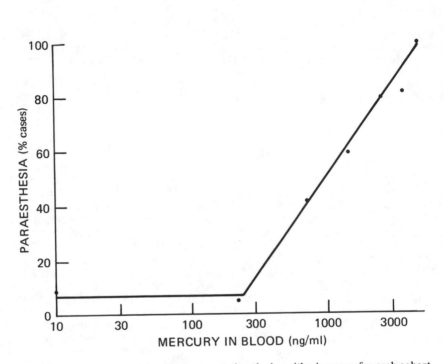

The mean blood concentrations are computed as the logarithmic means for each cohort in their table. The line connecting the first two points was assumed to be horizontal. The line connecting the other points was computed by least squares linear regression analysis.

SOURCE: Adapted from Bakir et al. (1973). Science 181:230-241. Copyright 1973 by the American Association for the Advancement of Science.

FIGURE 6.1 The frequency of paraesthesia as a function of the concentration of mercury in blood, 65 days after cessation of exposure.

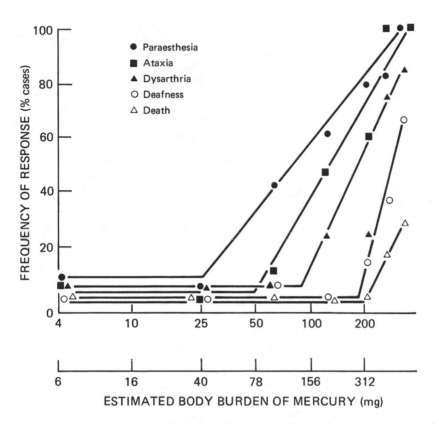

Both scales of the abscissa refer to body burdens of methylmercury (mg) at the cessation of exposure. The two scales represented different methods of calculating the body burden as discussed in the text.

SOURCE: Adapted from Bakir et al. (1973). Science 181:230-241. Copyright 1973 by the American Association for the Advancement of Science.

FIGURE 6.2 The relationship between frequency of signs and symptoms and the estimated body burden of methylmercury.

(Figure 6.2). Additional Iraqi data reported by
Kazantzis (1976) agree with these dose-response
estimates. In the exposed Iraqi population 31 percent
of women who had apparent symptoms and/or signs
attributable to methylmercury poisoning had maximum hair
mercury levels less than 100,000 ng/g. The authors
suggest that there is a category of methylmercury
poisoning in which symptoms such as paraesthesia,
headaches, persistant pain, and weakness of the limbs
predominate, with little or no evidence of neurological
damage on clinical examination, and that these effects
may occur with blood mercury levels well below 400
ng/ml. The estimate of the mean body burden that
results in the appearance of the first observable effect
(paraesthesia) in the exposed Iraqi population agrees
surprisingly well with the findings reported earlier by
the Swedish expert committee from analysis of data from
the Japanese outbreaks. They estimated a minimum body
burden of approximately 30 mg of mercury in
methylmercury form (70-kg person) to be associated with
the onset of the earliest signs and symptoms of
methylmercury poisoning (Berglund et al. 1971).

Assumptions and Reliability of the Estimates

These limits for detecting the effects of
methylmercury should not be equated with threshold
levels, since other more subtle effects may not be
observed with current techniques. The mean adult body
burden at which methylmercury begins to cause damage to
the nervous system can be assumed to be lower, perhaps
considerably lower, than current limits of detection of
minimal effect (paraesthesia). Limits of individual
variability in the population are approximately
reflected by the slope of the relationship between
response and body burden. For paraesthesia, individuals
in the Iraqi population vary by approximately a factor
of 10 (Bakir et al. 1973). The different slopes of the
other dose-response relationships shown in Figure 6.2
suggest that the extent of variation among individuals
differs for different signs or symptoms.
The reliability of these estimates of dose-response
and allowable daily intake depend upon the following
assumptions:

• nearly 100 percent absorption of methylmercury
compounds among individuals in the exposed population,
• turnover of methylmercury among various bodily
compartments, including the nervous system, that is
considerably faster than the elimination of
methylmercury from the body as a whole (implying a

constant ratio between the concentration of
methylmercury in the critical organ and in other
tissues); and
- elimination of methylmercury from the body
following a single exponential function (i.e., with
long-term exposure the steady state body burden in mg[B] is:
$B = T \times a/\ln 2$, where T = total body elimination
half-time, and a = daily dose of methylmercury in mg Hg
to a 50-kg person (Task Group on Metal Accumulation
1973).

Recent observations of the exposed Iraqi population
have documented variation in biological half-times among
exposed individuals. Using sequential hair analyses,
Al-Shahristani and Shibab (1974) have reported that
though most adults have half-times of approximately 70
days, a significant number of individuals may have
elimination half-times as long as 120 days. Since
elimination half-time has been shown to be directly
related to the cumulative body burden as a function of
the duration of exposure, individual variations in
excretion are important in estimating exposure risks.
Recently, Nordberg and Strangert (1976) incorporated
these data into a more complex metabolic formulation for
population risk evaluation. Their model estimates
directly the probability of observable toxicity among
individuals in an exposed population and does not
require the use of a "safety factor." Risk estimates
that use this method support the adequacy of the
earlier, Swedish expert group estimate of an acceptable
daily intake (ADI) of 30 µg of mercury as methylmercury
per day for a 70-kg person (Berglund et al. 1971). The
model specifically predicts that for this ADI there
would be fewer than 1 in 5000 exposed adult individuals
who would have observable toxicity.

Human Populations with
Large Intake of Freshwater Fish

Sweden

Birke et al. (1972) and Skerfving (1974) reported
observations on a total of nearly 200 individuals with
levels of methylmercury in the blood elevated by regular
consumption of fish taken from various freshwater and
coastal areas of Sweden. The fish contained mercury
levels of 0.3 to 7.0 µg/g, and human blood mercury
levels as high as 600 ng/ml were recorded. Thirty
percent of the subjects were examined in detail for
signs of mercury poisoning, and the 3 percent with the
highest fish consumption were intensively examined in

hospitals. No definitive signs or symptoms of methylmercury poisoning were observed.

Canada

Elevated levels of mercury in fish from Lake St. Clair and lakes in northwest Ontario were first detected in 1969 (Fimreite 1970). Since then, methylmercury concentrations significantly above the Canadian guideline (0.5 μg/g) have been documented widely in fish taken from lakes and waterways throughout northwest Ontario and northwest Quebec. Some of these waterways are known to have been polluted with mercury released from industrial sources, but others are not known to have been directly contaminated.

During 1975 residents of the White Dog and Grassy Narrows Reserves (Ontario) were surveyed to determine the extent of exposure to mercury through ingestion of large quantities of freshwater fish and to assess whether such exposure may pose a health hazard. Levels of mercury in the blood were observed to range widely, from less than 5 to 330 ng/ml (Clarkson 1975). Adult male guides at fishing camps had the highest levels. Their wives also tended to have higher blood levels than other women on the reserves. A newborn infant with a hair concentration of 30,000 ng/g, equivalent to a blood concentration of approximately 120 ng/ml, was also described.

During the summer of 1975, a Japanese team invited by the Canadian National Indian Brotherhood visited Grassy Narrows and White Dog Reserves and reported that 37 of 89 people examined had signs and symptoms of methylmercury poisoning (Harada et al. 1976).

Similar comprehensive surveys in northwest Quebec are under way and have documented significantly elevated blood and hair mercury levels in native Canadians known to consume large quantities of fish from freshwater lakes.

A Canadian clinical medical team headed by A. Barbeau examined native Canadian populations in northwest Quebec with significantly elevated blood mercury concentrations and reported that at least 6, and possibly 25, of 49 people examined had signs and symptoms of methylmercury intoxication (Barbeau et al. 1976). Approximately 80 percent of the individuals examined were reported to have had blood mercury levels above 50 ng/ml at the time of neurological examination.

The validity of these clinical evaluations by the Canadian and Japanese clinical teams has been questioned. Possible nutritional deficiencies and high alcohol intake by some of the individuals reported to be poisoned by methylmercury complicate the interpretation of clinical observations. A general consensus as to whether methylmercury intake from fish ingestion has resulted in neurological damage in Canadian native populations must await further studies, particularly histopathologic examinations of brain tissues from people documented to have significantly increased methylmercury body burdens, and properly designed epidemiological population surveys.

Human Populations with·
Large Intake of Ocean Fish

Eighty-eight tuna fishermen in American Samoa have been found to have an average blood methylmercury level of 64 ng/ml (range, 5 to 265 ng/ml) through eating large quantities of ocean fish (Marsh et al. 1974, Clarkson and Marsh 1976). In two villages in northern Peru, 186 inhabitants who had a high dietary intake of marine fish were observed to have a mean blood methylmercury concentration of 82 ng/ml (range, 11 to 275 ng/ml) (Turner et. al. 1974). Careful clinical neurological examination of these individuals revealed no evidence of clinical methylmercury intoxication in either the Peruvian or the Samoan populations. Astier-Dumas and Cumont (1975) and Kaku et al. (1975) have reported similar results in France and Japan, respectively. It should be noted that there were no reported observations of prenatally exposed individuals in these populations.

TOXIC EFFECTS OF METHYLMERCURY ON
DEVELOPING FETUSES AND INFANTS

It has become increasingly obvious that a life-cycle perspective is necessary for appropriate evaluation of potential adverse effects of environmental agents. All stages of the life cycle must be examined for sensitivity to exposure, and all stages subsequent to the exposure must be assessed to determine acute as well as possible delayed responses. Certain effects may become observable only after a latent period when the organism's adaptive capacities are stressed by other adverse environmental factors, or when aging processes reduce particular systems to smaller safety margins in terms of ranges of adaptive response capacities.

The critical organ concentration may differ for different stages of the human life cycle. The developing fetal (and newborn) brain may be the most sensitive organ (i.e., critical organ) in terms of human methylmercury toxicity (Berglund et al. 1971). Studies in animal models, as well as comparative qualitative observations of effects on prenatally exposed infants and mothers, support the notion that methylmercury is more toxic to the fetus than to the adult animal (Berglund et al. 1971, Spyker et al. 1972, Su and Okita 1976a).

During the Japanese Minamata outbreak, 23 infants with severe psychomotor signs of brain damage were described. They were born to mothers who had consumed fish taken from waters known to be heavily contaminated with effluent containing methylmercury from a nearby factory. In contrast to the apparent brain damage of their prenatally exposed infants, these mothers were reported to lack symptoms or signs of methylmercury poisoning other than mild paraesthesia. Thus, it was concluded that methylmercury crossed the placenta and that the fetal brain was much more sensitive than the adult brain (Harada 1968). However, no documentation of dose, fetal or maternal blood, or brain levels of methylmercury was possible in the Japanese cases. A similar observation of apparent fetal brain sensitivity was more recently reported following maternal ingestion of pork contaminated with methylmercury during the third to sixth months of pregnancy (Curley et al. 1971, Snyder 1971, Pierce et al. 1972).

In the large Iraqi outbreak of methylmercury poisoning, it has been possible to document prenatal transplacental exposure by direct measurement of maternal and fetal index media. Thus, for the first time it is possible to determine prenatal and early postnatal dose-response relationships. In all but one of the infant-mother pairs studied by Amin-Zaki et al. (1974, 1976), the infant's blood mercury level was higher than the mother's during the first few months of life. Ongoing sequential analyses of mercury in hair samples obtained from a larger group of women exposed during pregnancy should recapitulate exposure and permit estimation of the maternal and fetal dose.

Preliminary results on 29 children exposed in utero and examined at 3 and 4 1/2 years of age strongly suggest that the fetal brain is the critical organ in the exposed pregnant female. A clear difference was demonstrated by the group of children who were exposed in utero to maternal body burdens of methylmercury reflected by peak maternal hair concentrations of 112,000 to 384,000 ng/g (Figure 6.3; Marsh et al. [In press]). Clinical evidence of detectable fetal brain damage was observed when the peak maternal hair mercury concentration rose above 100,000 ng/g during pregnancy. A hair mercury concentration of 100,000 ng/g is estimated to be equivalent to a blood mercury concentration of approximately 400 ng/ml, assuming a

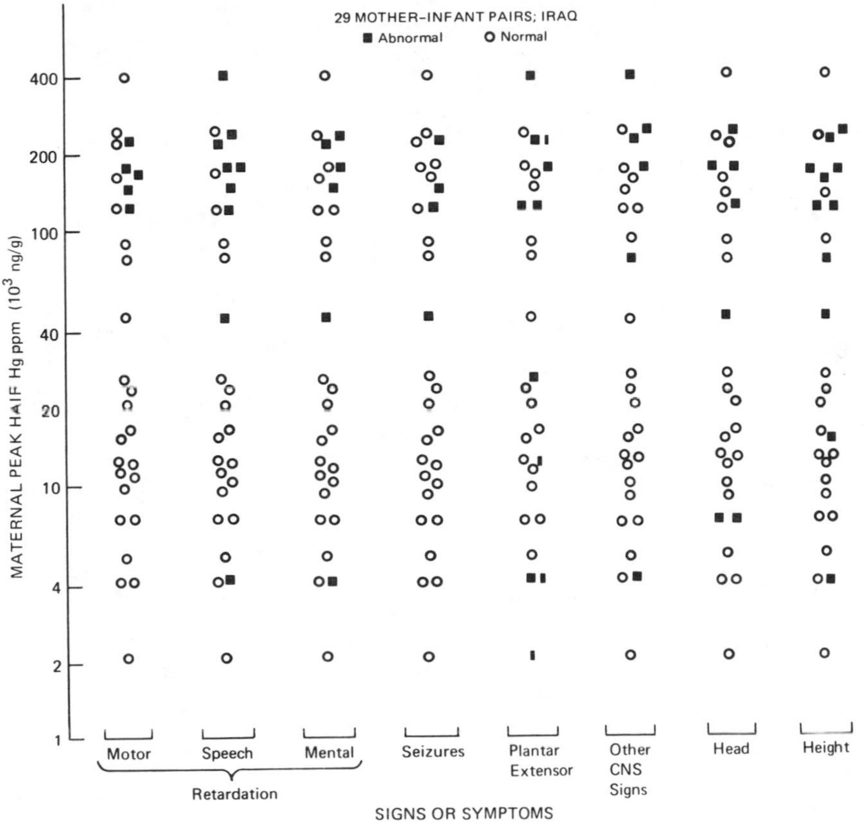

SOURCE: Marsh et al. (in press).

FIGURE 6.3 Signs or symptoms of mercury poisoning in children exposed *in utero* as a function of maternal hair mercury concentration (children were examined at ages 3 and 4.5 years).

hair:blood ratio of 250. Since only 29 infants were examined in this series, it is to be expected that the distribution of infants in a larger exposed population would contain individuals who would be affected at considerably lower maternal hair and blood concentrations. The conclusions of the international group of experts who authored the WHO (1977) report, Environmental Health Criteria 1, Mercury, concerning the concentrations of total mercury in indicator media associated with the earliest effects in the most sensitive group in the population are listed below.

Blood (ng/ml)	Hair (µg/g)	Equivalent long-term daily intake of mercury as methylmercury[a] (µg/kg body weight)
200-500	50-125	3-7

[a] A Japanese group has recently concluded that a daily intake of mercury of 5µg/kg is the "minimal toxic dose," following a 10-yr follow-up of the study of the Minamata outbreak (Research Committee on Minamata Disease 1975).

The WHO Task Group also urged that their conclusions not be considered independently of the section of their report entitled "Effects of Mercury on Man--Epidemiological and Clinical Studies."

EFFECTS OF METHYLMERCURY ON GENETIC AND REPRODUCTIVE PROCESSES[2]

In rapidly dividing Alium cepa (onion) root cells, methylmercury has been shown to interfere with normal chromosome segregation by disrupting the mitotic spindle function (chromosome segregation) and causing c-mitosis at an exceedingly low concentration, 50 ng/ml (2.5×10^{-7} M), 1000 times lower than the colchicine concentration required to produce similar mitotic arrest of cell division (Ramel 1967). Significant c-mitotic effects were also observed in Vicia faba exposed to 20 ng/ml methylmercury (1×10^{-7} M) (Ramel 1972), and in cultured human leukocytes exposed to 200 to 400 ng/ml (1 to 2×10^{-6} M) (Fiskesjo 1970). These findings have recently been confirmed in methylmercury exposed Vicia faba and Tradescantia root tips (Ahmed and Grant 1972).
C-Meiotic effects of methylmercury reported in Tradescantia and sex chromosome nondisjunction in female Drosophila melanogaster fed methylmercury document the

potential for interference with the meiotic cell cycle
as well, resulting in viable aneuploid offspring (Ramel
1972). Meiotic ova from random-bred Swiss/Webster mice
cultured in vitro for 5 to 14 hours in the presence of
mercuric acetate (MA) or dimethylmercury (DM) were
compared with control ova (not exposed to mercury) in
terms of effects on first and second metaphase meiotic
figures. Dose-related effects were seen for both
compounds. Fifty thousand ng MA/ml and 25,000 ng DM/ml
completely prevented cell divisions. Twenty-five
thousand ng MA/ml and 10,000 ng DM/ml severely disrupted
meiotic progression and resulted in second meiotic
metaphase abnormalities (Jagiello and Lin 1973).

Frolen and Ramel (Ramel 1972) reported that male CBA
mice treated with a single intraperitoneal dose of 3
mg/kg methylmercury dicyandiamide showed no detectable
dominant lethal effects when mated weekly during the
ensuing 6 weeks, as indicated by incidences of living
embryos and dead implantations. However, they did
observe a significant reduction in numbers of fertile
matings compared with controls. Khera (1973) treated
rats and mice orally with methylmercury chloride, 0,
1.0, 2.5, and 5.0 mg/kg/day for 7 days. In rats he
found a significant dose-related reduction in mean
litter size at all doses during days 5 to 20 post-
treatment, which was attributed to preimplantation
losses. Male Swiss/Webster mice were dosed orally each
day for 5 or 7 consecutive days with doses up to 5
mg/kg. With serial matings a slight, but not
statistically significant, reduction in the average
number of viable embryos was observed as a result of
preimplantation losses. Lee and Dixon (1975) reported a
reduction in male fertility of mice given a single
intraperitoneal dose of 1 mg/kg methylmercury hydroxide.
Fertility profiles from serial matings suggested an
effect on spermatogonial cells and premeiotic
spermatocytes.

Recently, Suter (1975) has also reported small but
significant fertility effects in mouse dominant lethal
and reproduction capacity studies. Male mice were
injected intraperitoneally with 10 mg/kg methylmercuric
hydroxide. No fertility effects were seen in serial
matings of treated (101 × C_3H)F_1 male mice during a 48-
day period. However, (SEC × $C_{57}Bl$)F_1 males showed
significant small reductions in the average number of
total and living implants during the early post-
treatment interval (~7.5 days). A significant small
reduction in numbers of progeny produced and in numbers
of living embryos was also seen in similarly treated
females mated within the first 4.5 days after treatment.

Ramel (1972) observed that methylmercury causes
chromosome breaks in plant root cells. Fiskesjo (1970)

found no chromosome breakage in human leukocytes treated
in vitro with methylmercury or dimethylmercury.
However, Skerfving et al. (1974) reported a
statistically significant correlation between
erythrocyte total mercury concentration and percentage
of cells with chromosome breaks or aneuploidy in
lymphocytes cultured from persons with elevated
erythrocyte methylmercury levels (range 13 to 1,100
ng/g) due to high intake of fish containing
methylmercury.

Using cultured mouse leukemic cells L5178Y exposed to
methylmercury in the range of 10^{-7} to 10^{-5} M, Nakazawa
et al. (1975) found retardation of cell multiplication,
depression of [^{3}H] thymidine and [^{3}H] uridine uptake, as
well as induction of single strand scissions of DNA.

Dimethylmercury has also been observed to cause single
strand DNA breaks in vivo in the slime mold Physarum
polycephalum. Interestingly, strains of Physarum
differing in geographical origin were found to have
widely different sensitivities to dimethylmercury damage
(Yatscoff and Cummins 1975).

A slight increase in the frequency of X-linked
recessive lethals ("point mutations") in Drosophila
after exposure to methylmercury in food was reported by
Ramel (1972); and recently, more significant increases
in X-linked recessive lethals following exposure of
Drosophila to ethyl mercury-p-toluene sulphanilamide
have been observed (Mathew and Al-Doori 1976).

Nakai and Machida (1973) were unable to detect nuclear
nonsense and frameshift reversion mutations in
Saccharomyces cerevisiae exposed to methylmercury, but
they did observe induction of cytoplasmic petite
mutations.

No assessment of mutagenicity in prokaryote systems
such as the Ames salmonella assay has yet been made
owing to technical difficulties of working with metals.

In summary, data concerning possible genetic and
reproductive effects of mercury compounds is meager and
somewhat contradictory. Methylmercury has been shown to
be a relatively weak mutagen in Drosophila. It can
interfere with mitotic and meiotic chromosome
segregation in plants and animals, and in some rodent
species it has been shown to cause reduced fertility.
Gonadal exposure has not been assessed in exposed human
individuals, and there are very few data from other
mammals. Exposure to methylmercury causes chromosomal
abnormalities in rapidly growing plant root cells, and
has been reported to cause chromosomal damage in
lymphocytes cultured in vitro from methylmercury exposed
human individuals. The significance of these
observations for human health remains unclear.

While methylmercury appears not to be a strong mutagen, its effects on human reproduction and chromosomes require further observations before the hazards can be more accurately evaluated.

TERATOGENIC EFFECTS OF METHYLMERCURY

Though significantly increased frequencies of congenital malformations such as cleft palate have been observed in different strains of mice exposed prenatally to methylmercury, no existing data on human exposure in utero implicate methylmercury as a teratogen (Spyker and Smithberg 1972, Su and Okita 1976b). It must, however, be emphasized that no adequate epidemiologic study of exposed human populations has been undertaken to assess possible teratogenic effects. The relatively small numbers of prenatally exposed individuals and logistical difficulties arising from geographic location make such studies extremely difficult to carry out. The degree of teratogenic risk from human prenatal exposure to methylmercury may thus remain undefined for the indefinite future.

CARCINOGENIC EFFECTS OF METHYLMERCURY

Little information has been published on animals and none on man with which to assess the potential carcinogenic hazard of exposure to mercury compounds. The carcinogenic potential of exposure to methylmercury appears to be low, but the available data are inadequate to make a reliable assessment.

In rats, sarcomas developed in areas in direct contact with metallic mercury injected intraperitoneally, but no metastases were observed (Druckery et al. 1957). Schroeder and Mitchener (1975) reported no significant difference in tumor frequencies between control, unexposed, random-bred white Swiss mice and mice with lifetime exposures to methylmercury in drinking water (5000 ng/ml × 70 days, then 1000 ng/ml subsequently). It has also been shown that prolonged exposure of mice to 1000 or 10,000 ng/g methylmercury in their feed did not alter the course of neoplasia following innoculation with Rauscher leukemia virus (Koller 1975).

Lifetime exposure studies have been done in both sexes of only one species, the mouse. Current recommendations suggest the use of two rodent species for carcinogenicity testing. Since transplacental exposure can occur, exposure should begin before conception, and the offspring should continue to be exposed for life (NRC 1975).

SUMMARY AND CONCLUSIONS

Though data about the methylmercury content of marine and freshwater fish and populations have accumulated rapidly during the past few years, several gaps remain in the information required to assess accurately the hazards of human dietary exposure to methylmercury in fish.

With the possible exception of certain Canadian native populations exposed to methylmercury through eating large quantities of freshwater fish, no human poisoning has yet been documented to result from eating ocean or freshwater fish from waters not directly contaminated by the anthropogenic release of mercury. In studies of acutely and chronically exposed adult human populations in Iraq, Samoa, Peru, and Sweden, no clinically affected adults were observed who had blood methylmercury levels below approximately 400 ng/ml. However, in the exposed Iraqi population approximately one-third of the women who had symptoms or signs attributable to methylmercury poisoning had maximum hair mercury levels less than 100,000 ng/g (estimated to be equivalent to blood levels less than 400 ng/ml).

However, it should be emphasized that the current clinical limits for detecting the effects of methylmercury in human populations should not be equated with threshold levels because other more subtle effects, such as behavioral or intellectual deficits, may not be detectable by the clinical procedures that were used.

It should also be noted that other environmental factors such as nutritional status and concurrent exposure to other intoxicants and infectious agents have not yet been evaluated for their additive or potentiating effects. The possible protective role of dietary selenium, particularly in ocean fish, remains to be determined (see discussion in Chapter 5).

The question of whether appropriate dose estimation should be based on peak concentrations in index media or concentrations integrated throughout the duration of exposure remains to be answered.

Prenatally exposed children in the Iraqi population are currently being observed but final conclusions will not be possible until these individuals have been observed through their reproductive years and aging. Observations made so far suggest that there is a high risk of clinically detectable fetal brain damage when maternal hair mercury concentrations rise above 100,000 ng/g (equivalent to a blood mercury concentration of approximately 400 ng/ml). Because only a small number of prenatally exposed infants has been observed, it is likely that studies of larger exposed populations would reveal a distribution with some infants affected at considerably lower maternal hair and blood concentrations. Continued careful evaluation of this very important cohort of prenatally exposed individuals will probably provide the most sensitive assessment of human methylmercury toxicity.

The human population is outbred, and genetic uniqueness of each individual (except for identical twins) can be assumed. Thus, variability among individuals is the rule, and though most individuals in a population will fall within a normal distribution in terms of dose response, there is a yet undetermined number who may have exceptional susceptibility to adverse effects of methylmercury owing to genetic and/or environmental factors. These individuals probably make up a small fraction of the human population, but we have as yet no means of estimating the numbers of people at special risk.

Relative susceptibility at different stages of the human life cycle, especially the developing organism (organogenesis through puberty), and sensitivity of aged or disabled individuals have not yet been adequately assessed. It is prudent to retain a considerable margin of safety in the exposure limits recommended for pregnant women as well as other potentially susceptible individuals. Until more definitive evaluations of the exposed native Canadian population and the prenatally and perinatally exposed Iraqi populations have been completed, the guidelines concerning human exposure to methylmercury suggested in the WHO document (1976b), Environmental Health Criteria 1, Mercury, should be adhered to.

NOTES

1 For more detailed information concerning toxic human health effects due to methylmercury exposure the reader is referred to the extensive reviews and summaries to be found in Berglund et al. (1971), U.K. Department of the Environment (1976), WHO (1976a, 1976b), and Tsubaki and Irukayama (1977).

2 The literature describing possible genetic, reproductive, teratogenic, and carcinogenic effects has been collated and described in greater detail than the toxic effects literature since there is no recent summary description available in the literature.

REFERENCES

Abbott, D.C. and J.O. Tatton (1970) Pesticide residues
    in the total diet in England and Wales, 1966-1967. IV.
    Mercury content of the total diet. Pesticide Science
    1:99-100.
Abramovskiy, B.P., Yu. A. Anokhin, V.A. Ionov, E.M.
    Nazarov, and A. Kh. Ostromogil'skiy (1975) Global
    balance and maximum permissible mercury emissions into
    the atmosphere. Pages 14-21, Second Joint
    U.S./U.S.S.R. Symposium on the Comprehensive Analysis
    of the Environment. October 21-26, 1975. Honolulu:
    U.S. Environmental Protection Agency.
Adin, A. and W. Espenson (1971) Kinetics for methyl-
    transfer to mercury. Chemical Communications 13:653-
    654.
Ahmed, M. and W.F. Grant (1972) Cytological effects of
    the mercurial fungicide Panogen 15 on tradescantia and
    Vicia faba root tips. Mutation Research 14:391-396.
Aho, I. (1968) The occurrence of mercury in Aland pike.
    Huso Biologiska Station Meddelande (Finland) 13:5-13.
Al-Shahristani, H. and K.M. Shibab (1974) Variation of
    biological half-life of methylmercury in man. Archives
    of Environmental Health 28:342-344.
Amend, D.F., W.T. Yasutake, and R. Morgan (1969) Some
    factors influencing the susceptibility of rainbow
    trout to the acute toxicity of an ethylmercuric
    phosphate formulation (Timsan). Transactions of the
    American Fisheries Society 98:419-425.
Amin-Zaki, L., S. Elhassani, M.A. Majeed, T.W. Clarkson,
    R.A. Doherty, and M.R. Greenwood (1974) Intrauterine
    methylmercury poisoning in Iraq. Pediatrics 54:587-
    595.
Amin-Zaki, L., S. Elhassani, M.A. Majeed, T.W. Clarkson,
    R.A. Doherty, M.R. Greenwood, and T. Giovanoli-
    Jakubczak (1976) Perinatal methylmercury poisoning in
    Iraq. American Journal of Diseases of Children
    130:1070-1076.
Anas, R.E. (1974) Heavy metals in the northern fur seal,
    Callorhinus ursinus, and harbor seal, Phoca vitulina
    richardi. Fisheries Bulletin 72:133-137.

107

Anderson, W.L. and P.L. Stewart (1971) Incidence of mercury in Illinois pheasants. Transactions of the Illinois State Academy of Science 64:237-241. (Chemical Abstracts 77:15187e.)

Andersson, A. (1967) Kvicksilvret i marken. Grundforbattring 20:95-105.

Andren, A.W. and R.C. Harriss (1975) Observations on association between mercury and organic matter dissolved in natural waters. Geochimica et Cosmochimica Acta 39:1253-1257.

Anfalt, D., D. Dyrssen, E. Ivanova, and D. Jagner (1968) State of divalent mercury in natural waters. Svensk Kemisk Tidskrift 80:340-342.

Annett, C.S., F.M. D'Itri, J.R. Ford, and H.H. Prince (1975) Mercury in fish and waterfowl from Ball Lake, Ontario. Journal of Environmental Quality 4:219-222.

Anonymous (1971) Mercury in whales. Marine Pollution Bulletin 2:68.

Astier-Dumas, M. and G. Cumont (1975) Weekly intake of fish and mercury levels in the blood and hair in France. Annales d'Hygiene de Langue Francaise - Medecine et Nutrition 1:135-139.

Aston, D., D. Bruty, R. Chester, and J.P. Riley (1972) The distribution of mercury in the N. Atlantic deep-sea sediments. Nature (Physical Science) 237:125.

Bache, C.A., W.H. Gutenmann, and D.I. Lisk (1971) Residues of total mercury and methylmercury salts in lake trout as a function of age. Science 172:951-952.

Bache, C.A., W.H. Gutenmann, L.E. St. John, R.D. Sweet, H.H. Hatfield, and D.J. Lisk (1973) Mercury and methylmercury content of agricultural crops grown on soils treated with various mercury compounds. Journal of Agricultural and Food Chemistry 21:607-613.

Backstrom, J. (1969) Distribution studies of mercuric pesticides in quail and some freshwater fishes. Acta Pharmacologica et Toxicologica (Supplementum 3) 27:74-92.

Bails, J.D. (1972) Mercury in fish in the Great Lakes. Pages 31-37, Environmental Mercury Contamination, edited by R. Hartung and B.D. Dinman. Ann Arbor, Mich.: Ann Arbor Science Publishers.

Bakir, F., S.F. Damluji, L. Amin-Zaki, M. Murtadha, A. Khalidi, N.Y. Al-Rawi, S.T. Kriti, H.I. Dhahir, T.W. Clarkson, J.C. Smith, and R.A. Doherty (1973) Methylmercury poisoning in Iraq. Science 181:230-241.

Barbeau, A., A. Nantel, and F. Dorlot (1976) Etude sur les Effets Medicaux et Toxicologiques du Mercure Organique dans le Nord-Ouest Quebecois. Comite d'etude et d'intervention sur le mercure au Quebec. Quebec: Ministere des Affaires Sociales.

Barber, R.T., A. Vijayakumar, and F.A. Cross (1972) Mercury concentrations in recent and ninety-year-old benthopelagic fish. Science 178:636-639.

Barnes, H. and F.A. Stanbury (1948) The toxic action of copper and mercury salts both separately and when mixed on the harpactacid copepod, Nitocra spinipes (Boeck). Journal of Experimental Biology 25:270-275.

Beasley, T.M. (1971) Mercury in selected fish protein concentrates. Environmental Science and Technology 5:634-635.

Belisle, A.A., W.L. Reichel, L.N. Locke, T.G. Lamont, B.M. Mulhern, R.M. Prouty, R.B. DeWolf, and E. Cromartie (1972) Residues of organochlorine pesticides, polychlorinated biphenyls, and mercury and autopsy data for bald eagles, 1969 and 1970. Pesticides Monitoring Journal 6:133-138.

Berg, W., A. Johnels, B. Sjostrand, and T. Westermark (1966) Mercury content in feathers of Swedish birds from the past 100 years. Oikos 17:71-83.

Berglund, F., M. Berlin, G. Birke, R. Cedarlof, U. von Euler, L. Friberg, B. Holmstedt, E. Jonsson, K.G. Luning, C. Ramel, S. Skerfving, A. Swensson, and S. Tejning (1971) Methylmercury in fish, A toxicologic-epidemiologic evaluation of risks. Report from an Expert Group. Nordisk Hygienisk Tidskrift (Supplementum 4).

Billen, G. (1973) Etude de l'ecometabolisme du mercure dans un milieu d'eau douce. Hydrobiological Bulletin 7:60-68.

Billen, G. and R. Wollast (1973) Transformations biologiques du mercure dans les sediments de la Sambre. Pages 191-232, Rapport de synthese, project Sambre. Journees d'etude des 27 et 28 novembre, 1972. CIPS.

Billings, C.E., A.M. Sacco, W.R. Matson, R.M. Griffin, W.R. Coniglio, and R.A. Harley (1973) Mercury balance on a large, pulverized coal-fired furnace. Journal of the Air Pollution Control Association 23:773-777.

Birke, G., A.G. Johnels, L.O. Plantin, B. Sjostrand, S. Skerfving, and T. Westermark (1972) Humans exposed to methylmercury through fish consumption. Archives of Environmental Health 25:77-91.

Bishop, J.N. and D. Boomer (1974) The relationship between mercury and selenium in freshwater fish. Presented at the 15th Great Lakes Conference, April 4-8, 1974, Albany, New York. Sponsored by the International Joint Commission. Unpublished.

Bishop, J.N. and B.P. Neary (1974) The form of mercury in freshwater fish. Pages III-25-29, Proceedings International Conference on Transport of Persistent Chemicals in Aquatic Ecosystems, National Research Council of Canada. Ottawa, Canada.

Bishop, J.N. and B.P. Neary (1976) Mercury Levels in Fish from Northwestern Ontario, 1970-1975. Inorganic Trace Contaminants Section, Ministry of the Environment, Rexdale, Ontario: Laboratory Services Branch.

Bishop, J.N. and B.P. Neary (1977) The decline in the mercury concentation from Lake St. Clair, 1970-1976. Report No. AQS 77-3, Rexdale, Ontario, Canada: Ministry of the Environment.

Bisogni, J.J. and A.W. Lawrence (1973) Kinetics of Microbially Mediated Methylation of Mercury in Aerobic and Anaerobic Aquatic Environments. Report to OWRR, Department of the Interior. Technical Report No. 63. Ithaca, N.Y.: Cornell University Water Resources and Marine Sciences Center.

Bisogni, J.J. and A.W. Lawrence (1975a) Kinetics of mercury methylation in aerobic and anaerobic environments. Journal of the Water Pollution Control Federation 47:135-152.

Bisogni, J.J. and A.W. Lawrence (1975b) Metabolic cycles for toxic elements in the environment: A study of kinetics and mechanism (J.M. Wood). Pages 113-115, Heavy Metals in the Aquatic Environment, edited by P.A. Krenkel. Oxford: Pergamon Press.

Blaylock, B.G. and J.W. Huckabee (1974) The uptake of methyl mercury by aquatic biota. Pages III-73-74, Proceedings of the International Conference on Transport of Persistent Chemicals in Aquatic Ecosystems, National Research Council of Canada. Ottawa, Canada.

Bligh, E.G. (1970) Mercury and the contamination of freshwater fish. Manuscript Report No. 1088. Winnipeg, Canada: Fisheries Research Board of Canada, Freshwater Institute.

Bligh, E.G. (1971) Mercury levels in Canadian fish. Pages 73-90, Proceedings of the Symposium on Mercury in Man's Environment, 15-16 February 1971. Ottawa, Canada: Royal Society of Canada.

Blus, L.J., A.A. Belisle, and R.M. Prouty (1974) Relations of the brown pelican to certain environmental pollutants. Pesticides Monitoring Journal 7:181-194.

Boetius, J. (1960) Lethal action of mercuric chloride and phenylmercuric acetate on fishes. Meddelelser fra Danmarks Fiskeri-og Havundersogelser 3:93-115. (Biological Abstracts 37:16971.)

Boney, A.D. (1971) Sublethal effects of mercury on marine algae. Marine Pollution Bulletin 2:69-71.

Boney, A.D. and E.D.S. Corner (1959) Application of toxic agents in the study of the ecological resistance of intertidal red algae. Journal of the Marine

Biological Association of the United Kingdom 38:267–275.

Boney, A.D., E.D.S. Corner, and B.W.P. Sparrow (1959) The effects of various poisons of the growth and viability of sporelings of the red alga Plumaria elegans (Bonnem.) Schm. Biochemical Pharmacology 2:37–49.

Booer, J.R. (1944) The behaviour of mercury compounds in soils. Annals of Applied Biology 31:340–359.

Borg, K., K. Erne, E. Hanko, and H. Wanntorp (1970) Experimental secondary methylmercury poisoning in the goshawk (Accipiter g. gentilis L.). Environmental Pollution 1:91–104.

Borg, K., H. Wanntorp, K. Erne, and E. Hanko (1966) Mercury poisoning in Swedish wildlife. Journal of Applied Ecology 3 (Suppl.):171–172.

Borg, K., H. Wanntrop, K. Erne, and E. Hanko (1969) Alkylmercury poisoning in terrestrial Swedish wildlife. Viltrevy 6:301–379.

Breck, D.W., F. Chields, E.R. Norberg, and J.E. Cline (1973) An analysis of mercury residues in Idaho pheasants. Pages 186–198, Mercury in the Western Environment, edited by D.R. Buhler. Corvallis, Oreg.: Continuing Education Publications.

Brouzes, R.J.P., R.A.N. McLean, and G.H. Tomlinson (1977) Mercury – The Link Between pH of Natural Waters and the Mercury Content of Fish. Research Report. Montreal, Quebec: Domtar Ltd. Research Center.

Buhler, D.R., R.R. Claeys, and H.J. Rayner (1973) Seasonal variations in mercury contents of Oregon pheasants. Pages 199–211, Mercury in the Western Environment, edited by D.R. Buhler. Corvallis, Oreg.: Continuing Education Publications.

Burk, R.F., K.A. Foster, P.M. Greenfield, and K.W. Kiker (1974) Binding of simultaneously administered inorganic selenium and mercury to a rat plasma protein. Proceedings of the Society for Experimental Biology and Medicine 145:782–785.

Burrows, W.D., K.I. Taimi, and P.A. Krenkel (1974) The uptake and loss of methylmercury by freshwater fish. Pages 283–288, Proceedings Congreso Internactional del Mercurio, Tomo II. Barcelona, Spain.

Carr, R.A., J.B. Hoover, and P.E. Wilkniss (1972) Cold-vapor atomic absorption analysis for mercury in the Greenland Sea. Deep Sea Research 19:747–752.

Carr, R.A. and P.E. Wilkniss (1973) Mercury in the Greenland Ice Sheet: Further data. Science 181:843–844.

Chamberlain, A.C. (1960) Aspects of the deposition of radioactive gases and particles. Pages 63–88, Proceedings of the Conference on Aerodynamic Capture of Particles, B.C.U.R.A. Leatherhead, Surrey, 1960, edited by E.G. Richardson. London: Pergamon Press.

Chau, Y.K. and H. Saitoh (1973) Determination of methylmercury in lake water. International Journal of Environmental Analytical Chemistry 3:133-139.

Clark, G.L. (1947) Poisoning and recovery in barnacles and muscles. Biological Bulletin (Woods Hole, Mass.) 92:73-91.

Clarkson, T.W. (1975) Exposure to methylmercury in Grassy Narrows and White Dog Reserves: an Interim Report. Medical Services Branch, Department of Health and Welfare, Canadian Federal Government.

Clarkson, T.W. and D.O. Marsh (1976) The toxicity of methylmercury in man: dose-response relationships in adult populations. Pages 246-261, Effects and Dose-Response Relationships of Toxic Metals: Proceedings from an international meeting organized by the Subcommittee on the Toxicology of Metals of the Permanent Commission and International Association on Occupational Health, Tokyo, Nov. 18-23, 1974, edited by G.F. Nordberg. Amsterdam: Elsevier Scientific Pub. Co.

Connor, P.M. (1972) Acute toxicity of heavy metals to some marine larvae. Marine Pollution Bulletin 3:190-192.

Corneliussen, P.E. (1969) Residues in food and feed. Pesticide residues in total diet samples (IV). Pesticides Monitoring Journal 2:140-152.

Corner, E.D.S. and F.H. Rigler (1958) The modes of action of toxic agents: III. Mercuric chloride and n-amylmercuric chloride on crustaceans. Journal of the Marine Biological Association of the United Kingdom 37:85-96.

Corner, E.D.S. and B.W. Sparrow (1957) The modes of action of toxic agents: II. Factors influencing the toxicities of mercury compounds to certain crustacea. Journal of the Marine Biological Association of the United Kingdom 36:459-472.

Cranston, R.E. and D.E. Buckley (1972) Mercury pathways in a river and estuary. Environmental Science and Technology 6:274-278.

Curley, A., V.A. Sedlak, E.F. Girling, R.E. Hawk, W.F. Barthel, P.E. Pierce, and W.H. Likosky (1971) Organic mercury identified as the cause of poisonings in humans and hogs. Science 172:65-67.

Dalgaard-Mikkelsen, S. (1969) Kviksolvforekomsten I miljoet I Danmark. (The occurrence of mercury in the Danish environment.) Nordisk Hygienisk Tidskrift 2/69 50:34-36.

de Frietas, A.S.W., S.U. Qadri, and B.E. Case (1974) Origins and fate of mercury compounds in fish. Pages III-31-36, Proceedings of the International Conference on Transport of Persistent Chemicals in Aquatic Ecosystems, National Research Council of Canada. Ottawa, Canada.

Delong, R.L., W.G. Gilmartin, and J.G. Simpson (1973) Premature births in California sea lions: Association with high organochlorine pollutant residue levels. Science 181:1168-1170.

Dencker, I. and A. Schutz (1971) Mercury content of food. Lakartidningen 68:4031-4033.

Desai-Greenway, P. and I.M. Price (1976) Mercury in Canadian Fish and Wildlife Used in the Diets of Native Peoples. Canadian Wildlife Service, Toxic Chemicals Division, Manuscript Report No. 35, Ottawa, Canada.

DeSimone, R.E., M.W. Penley, L. Charbonneau, S.G. Smith, J.M. Wood, H.A.O. Hill, J.M. Pratt, S. Ridsdale, and R.J.P. Williams (1973) The kinetics and mechanism of methyl and ethyl transfer to mercuric ion. Biochimica et Biophysica Acta 304:851-863.

Doi, R. and J. Ui (1975) The distribution of mercury in fish and its form of occurrence. Pages 197-221, Heavy Metals in the Aquatic Environment, edited by P.A. Krenkel. Oxford: Pergamon Press.

Dolar, S.G., D.R. Keeney, and G. Chesters (1971) Mercury accumulation by Myriophyllum spicatum L. Environmental Letters 1:191-198.

Druckery, H., H. Hamperl, and D. Schmahl (1957) Carcinogenic action of metallic mercury after intraperitoneal administration to rats. Zeitschrift fuer Krebsforchung 61:511-519.

Dustman, E.H., L.F. Stickel, and J.B. Elder (1972) Mercury in wild animals, Lake St. Clair, 1970. Pages 46-52, Environmental Mercury Contamination, edited by R. Hartung and B.D. Dinman. Ann Arbor, Mich.: Ann Arbor Science Publishers.

Dyrssen, D. and M. Wedborg (1974) Equilibrium calculations of the speciation of elements in sea water. Pages 181-195, The Sea: Ideas and Observations on Progress in the Study of Seas. Volume 5, Marine Chemistry, edited by E.D. Goldberg. New York: John Wiley & Sons.

Eades, J.F. (1966) Pesticide residues in the Irish environment. Nature 210:650-652.

Edelstam, C., A.G. Johnels, M. Olsson, and T. Westermark (1969) Ecological aspects of the mercury problem. Nordisk Hygienisk Tidskrift 50:14-28.

El-Begearmi, M.M., H.E. Ganther, and M.L. Sunde (1974) Effect of some sulfur amino acids, selenium, and arsenic on mercury toxicity using Japanese quail. Poultry Science 53(5):1921.

El-Begearmi, M.M., H.E. Ganther, and M.L. Sunde (1975) More evidence for a selenium arsenic interaction in modifying mercury toxicity. Poultry Science 54:1756-57. (Abstract only.)

El-Begearmi, M.M., H.E. Ganther, and M.L. Sunde (1976) Vitamin E decreases methylmercury toxicity. Poultry Science 55:2033. (Abstract only.)

El-Begearmi, M.M., C. Goudie, H.E. Ganther, and M.L. Sunde (1973) Attempts to quantitate the protective effect of selenium against mercury toxicity using Japanese quail. Federation Proceedings (Abst. No. 3756) 32:886.

Environment Canada (1971) High mercury levels found in ducks taken from the Wabigoon River, near Dryden, Ontario. News release 1-7161, September 10, Ottawa, Canada.

Environment Canada (1972) Enquete scientifique relative a la provenance et a la distribution du mercure dans l'environment du Nord-Quest Quebecois. Gouvernment du Canada, Gouvernment de la Province de Quebec.

Evans, R.J., J.D. Bails, and F.M. D'Itri (1972) Mercury levels in muscle tissues of preserved museum fish. Environmental Science and Technology 6:901-905.

Faber, R.A. and J.J. Hickey (1973) Eggshell thinning, chlorinated hydrocarbons, and mercury in inland aquatic bird eggs, 1969 and 1970. Pesticides Monitoring Journal 7:27-36.

Faber, R.A., R.W. Risenbrough, and H.M. Pratt (1972) Organochlorines and mercury in common egrets and great blue herons. Environmental Pollution 3:111-122.

Fang, S.C. (1973) Uptake and biotransformation of phenylmercuric acetate by aquatic organisms. Archives Environmental Contamination and Toxicology 1:18-26.

Food and Agriculture Organization/World Health Organization (FAO/WHO) (1968) 1967 Evaluation of some pesticide residues in food. Joint Meeting of the FAO Working Party of Experts and the WHO Expert Committee on Pesticide Residues. Rome, December 4-11, 1967. PL-CP/15 WHO/Food Add. 67 Accession #02778-67-MR. Geneva: World Health Organization.

Fimreite, N. (1970) Mercury uses in Canada and their possible hazards as sources of mercury contamination. Environmental Pollution, An International Journal 1:119-131.

Fimreite, N. (1974) Mercury contamination of aquatic birds in northwestern Ontario. Journal of Wildlife Management 38:120-131.

Fimreite, N., R.W. Fyfe, and J.A. Keith (1970) Mercury contamination of Canadian prairie seed eaters and their avian predators. Canadian Field Naturalist 84:269-276.

Fimreite, N., W.N. Holsworth, J.A. Keith, P.A. Pearce, and I.M. Gruchy (1971) Mercury in fish and fish-eating

birds near sites of industrial contamination in
Canada. Canadian Field Naturalist 85:211-220.

Fimreite, N. and L.M. Reynolds (1973) Mercury
contamination of fish in northwestern Ontario. Journal
of Wildlife Management 37:62-68.

Fiskesjo, G. (1970) The effect of two organic mercury
compounds on human leukocytes in vitro. Hereditas,
64:142-146.

Fitzgerald, W.F. and W.B. Lyons (1973) Organic mercury
compounds in coastal waters. Nature 242:452-453.

Foreback, C.C. (1973) Ph.D. Thesis. Tampa, Fla:
University of South Florida.

Forrester, C.R., K. Ketchem, and C.C. Wong (1972)
Mercury content of spiney dogfish in the Strait of
Georgia, British Columbia. Journal of the Fisheries
Research Board of Canada 29:1487-1490.

Freeman, H.C. and D.A. Horne (1973) Total mercury and
methylmercury content of the American eel (Anguilla
rostrata). Journal of the Fisheries Research Board of
Canada 30:454-455.

Froseth, J.A., R.C. Piper, and J.R. Carlson (1974)
Relationship of dietary selenium and oral
methylmercury to blood and tissue selenium
concentrations and deficiency-toxicity signs in swine.
Federation Proceedings (Abstr. No. 2543) 33:660.

Fujita, T. (1966) Mercury content in rice. Shiga
Kenritsu Eisei Kenkyushoho 7:14-15. Chemical Abstracts
73:86683x, 1970.

Furukawa, K., T. Suzuki, and K. Tonomura (1969)
Decomposition of organic mercurial compounds by
mercury-resistant bacteria. Agricultural and
Biological Chemistry 33:128-130.

Furukawa, K. and K. Tonomura (1971) Enzyme system
involved in the decomposition of phenyl mercuric
acetate by mercury-resistant Pseudomonas. Agricultural
and Biological Chemistry 35:604-610.

Furukawa, K. and K. Tonomura (1972a) Induction of
metallic mercury-releasing enzyme in mercury-resistant
Pseudomonas. Agricultural and Biological chemistry
36:2441-2448.

Furukawa, K. and K. Tonomura (1972b) Metallic mercury-
releasing enzyme in mercury-resistant Pseudomonas.
Agricultural and Biological Chemistry 36:217-226.

Ganther, H.E. (In press) Modification of methylmercury
toxicity and metabolism by selenium and Vitamin E:
Possible mechanisms, factors influencing
susceptibility to metal toxicity. In Proceedings of an
International Symposium on Factors Influencing
Susceptibility to Methyltoxicity, edited by L.
Friberg. Sponsored by Karolinska Institute in
Stockholm, Sweden, July 17-23, 1977. Environmental

Health Perspectives. Washington, D.C.: U.S. Department of Health, Education, and Welfare.

Ganther, H.E., C. Goudie, M.L. Sunde, M.J. Kopecky, P. Wagner, S.-H. Oh, and W.G. Hoekstra (1972) Selenium: Relation to decreased toxicity of methylmercury added to diets containing tuna. Science 175:1122-1124.

Ganther, H.E. and M.L. Sunde (1974) Effect of tuna fish and selenium on the toxicity of methylmercury: A progress report. Journal of Food Science 39:1-5.

Ganther, H.E., P.A. Wagner, M.L. Sunde, and W.G. Hoekstra (1973) Protective effects of selenium against heavy metal toxicities. Pages 247-252, Trace Substances in Environmental Health - VI. Proceedings of University of Missouri's 6th Annual Conference on Trace Substances in Environmental Health, June 13-15, 1975, edited by Delbert D. Hemphill. Columbia, Mo.: University of Missouri.

Garrels, R.M., F.T. MacKenzie, and C. Hunt (1973) Chemical Cycles and the Global Environment: Assessing Human Influences. Los Altos, Calif.: William Kaufman, Inc.

Gaskin, D.E., K. Ishida, and R. Frank (1972) Mercury in harbour porpoises (Phocoena phocoena) from the Bay of Fundy region. Journal of the Fisheries Research Board of Canada 29:1644-1646.

Gaskin, D.E., R. Frank, M. Holdrinet, K. Ishida, C.J. Walton, and M. Smith (1973) Mercury, DDT, and PCB in harbour seals (Phoca vitulina) from the Bay of Fundy and Gulf of Maine. Journal of the Fisheries Research Board of Canada 30:471-475.

Gerdes, R.A., J.E. Hardcastle, and K.T. Stabenow (1974) Mercury content of fresh fruit and vegetables. Chemosphere 3:13-18.

Gibbs, O.S., R. Shank, H. Pond, and G.H. Hansmann (1941) Absorption of externally applied ammoniated mercury. Archives of Dermatology and Syphilology 44:862-872. (Chemical Abstracts 36:16693.)

Gibbs, R.J., R. E. Jarosewich, and H.L. Windom (1974) Heavy metal concentrations in museum fish specimens: Effects of preservatives and time. Science 184:475-477.

Glooschenko, W.A. (1969) Accumulation of ^{203}Hg by the marine diatom Chaetoceros costatum. Journal of Phycology 5:224-226.

Goldwater, L.J. (1964) Occupational exposure to mercury. The Harben lectures. Journal of the Royal Institute of Public Health 27:279-301.

Goldwater, L.J. (1971) Mercury in the environment. Scientific American 224(5):15-21.

Gomez, M.I. (1972) Mercury Levels in Some Selected Foods and Evaluation of Assay Techniques. M.S. Thesis. East Lansing, Mich.: Michigan State University.

Greichus, Y.A., A. Greichus, and R.J. Emerick (1973)
Insecticides, polychlorinated biphenyls and mercury in
wild cormorants, pelicans, their eggs, food and
environment. Bulletin of Environmental Contamination
and Toxicology 9:321-328.

Griffith, W.H. (1973) Mercury contamination in
California's fish and wildlife. Pages 135-136, Mercury
in the Western Environment, edited by D.R. Buhler.
Corvallis, Oreg.: Continuing Education Publications.

Groth, D.H., L. Vignati, L. Lowry, G. Mackay, and H.E.
Stokinger (1973) Mutual antagonistic and synergistic
effects of inorganic selenium and mercury salts in
chronic experiments. Pages 187-189, Trace Substances
in Environmental Health - VI, Proceedings of
University of Missouri's 6th Annual Conference on
Trace Substances in Environmental Health, edited by
D.D. Hemphill. Columbia, Mo.: University of Missouri.

Gurba, J.B. (1970) Mercury situations in Alberta. Pages
53-73, Proceedings, 18th Annual Meeting and
Conference, Canada Agricultural Chemical Association.
Jasper, Alberta, Canada.

Hall, A.S., F.M. Teeny, L.G. Lewis, W.H. Hardman, and
E.J. Gauglitz, Jr. (1976a) Mercury in fish and
shellfish of the northeast Pacific. I. Pacific
halibut, Hippoglossus stenolepis. Fishery Bulletin
74(4):783-789.

Hall, R.A., E.G. Zook, G.M. Meaburn, T.L. Chambers, S.W.
Nealis, and J.J. Powell (1976b) Microconstituents
Resource Survey - Final Data Report. Southeastern
Utilization Research Center. College Park, Md.:
National Marine Fisheries Service.

Hamilton, A.L. (1971) Accumulation of mercury in fish
food organisms. Pages 73-90, Proceedings of the
Symposium on Mercury in Man's Environment, 15-16
February 1971. Ottawa, Canada: Royal Society of
Canada. (Included as part of E.G. Bligh; see reference
above.)

Hammerstrom, R.J., D.E. Hissong, F.C. Kopfler, J. Mayer,
E.F. McFarren, and B.H. Pringle (1972) Mercury in
drinking-water suplies. Journal of the American Water
Works Association 64:60-61.

Hammond, A.L. (1971) Mercury in the environment: natural
and human factors. Science 171:788-789.

Hannan, P.J. and C. Patouillet (1972) Effect of mercury
on algal growth rates. Biotechnology Bioengineering
14:93-101.

Hannerz, L. (1968) Experimental investigations on the
accumulation of mercury in water organisms. Report of
the Institute of Freshwater Research Drottningholm
48:120-176.

Harada, H., K. Ito, K. Ebato, M. Takeuchi, T. Amemiya,
H. Yamanobe, S. Suzuki and T. Totani (1975) Effect of

selenium on the toxicity of methylmercury. II.
Methylmercury and total mercury concentration of
organs in rats administered methylmercury, selenium,
and vitamin E. Annual Report of the Tokyo Metropolitan
Research Laboratory of Public Health 26:123-128.
(Chemical Abstracts 84:145588S.)

Harada, M., T. Fujino, T. Akagi, and S. Nishigaki (1976)
Epidemiological and clinical study and historical
background of mercury pollution in Indian reservations
in Northwestern Ontario, Canada. Bulletin of the
Institute of Constitutional Medicine, Kumamoto
University, XXVI (No. 3-4) 169-184.

Harada, Y. (1968) Clinical investigations on Minamata
Disease. Congenital (or fetal) Minamata Disease. Pages
92-117, Minamata Disease, edited by M. Kutsuna.
Kumamato, Japan: Kumamoto University Press.

Harlan, J.R. (1971) Mercury pollution survey. Sport
Fisheries Institute Bulletin No. 221:4-7.

Harris, E.J. and R.W. Karcher, Jr. (1972) Mercury: Its
historical presence in New York State fishes. Chemist
(NY)49:176-179.

Harriss, R.C., D.B. White, and R.B. Macfarlane (1970)
Mercury compounds reduce photosynthetis by plankton.
Science 170:736-737.

Hasselrot, T.B. (1968) Report of current field
investigations concerning the mercury content in fish,
bottom sediments and water. Report of the Institute of
Freshwater Research Drottningholm 48:102-111.

Hasselrot, T.B. and A. Gothberg (1974) The ways of
transport of mercury to fish. Pages III-37-47,
Proceedings of the International Conference on
Transport of Persistent Chemicals in Aquatic
Ecosystems. National Research Council of Canada,
Ottawa, Canada.

Heath, R.G. and S.A. Hill (1974) Nationwide
organochlorine and mercury residues in wings of adult
mallards and black ducks during the 1969-1970 hunting
season. Pesticides Monitoring Journal 7:153-164.

Heindryckx, R. et al. (1974) In Proceedings of the
International Symposium on the Problems of
Contamination of Man and His Environment by Mercury
and Cadmium, Luxembourg, 3-5 July, 1973. Luxembourg:
Commission of the European Communities.

Helminen, M., E. Karppanen, and J.I. Koivisto (1968)
Mercury in Finnish fresh water seals in 1967. Finsk
Veterinaertidskrift 74:87-89.

Hem, J.D. (1970) Chemical behaviour of mercury in
aqueous media. Pages 19-24, Mercury in the
Environment. U.S. Geological Survey Professional Paper
713. Washington, D.C.: U.S. Government Printing
Office.

Henderson, C., A. Inglis, and W.L. Johnson (1972)
  Mercury residues in fish 1969-1970 - national
  pesticide monitoring program. Pesticides Monitoring
  Journal 6:144-159.
Henderson, C. and W.E. Shanks (1973) Mercury
  concentrations in fish. Pages 45-58, Mercury in the
  Western Environment, edited by D.R. Buhler. Corvallis,
  Oreg.: Continuing Education Publications.
Henriksson, K., E. Karppanen, and M. Helminen (1969)
  Mercury in inland and marine seals. Nordisk Hygienisk
  Tidskrift 50:54-59.
Heppleston, P.B. and C.M. French (1972) Mercury and
  other metals in British seals. Nature 243:302-304.
Hill, C.H. (1974) Reversal of selenium toxicity in
  chicks by mercury, copper, cadmium. Journal of
  Nutrition 104:593-598.
Hill, H.A.O., J.M. Pratt, S. Ridsdale, F.R. Williams,
  and R.J.P. Williams (1970) Kinetics of substitution of
  co-ordinated carbanions in cobalt (111) corrinoids.
  Chemical Communications 6:341-342.
Holden, A.V. (1972) Present levels of mercury in man and
  his environment. Pages 143-168, Mercury Contamination
  in Man and His Environment. Vienna: International
  Atomic Energy Agency.
Holden, A.V. (1973a) International cooperative study of
  organochlorine and mercury residues in wildlife, 1969-
  1971. Pesticides Monitoring Journal 7:37-52.
Holden, A.V. (1973b) Mercury in fish and shellfish. A
  review. Journal of Food Technology 8:1-25.
Holt, G. (1969) Mercury residues in wild birds in Norway
  1965-1967. Nordisk Veterinaermedicin 21:105-114.
Huckabee, J.W., F.O. Cartan, and G.S. Kennington (1972)
  Distribution of mercury in pheasant muscle and
  feathers. Journal of Wildlife Management 36:1306-1309.
Hunter, W.R. (1949) The poisoning of Marinogammarus
  marinus by cupric sulphate and mercuric chloride.
  Journal of Experimental Biology 26:113-124.
Hussain, M. and E.L. Bleiler (1973) Mercury in
  Australian oysters. Marine Pollution Bulletin 4:44.
Imura, N., S.-K. Pan, and T. Ukita (1972) Methylation of
  inorganic mercury with liver homogenate of tuna fish.
  Chemosphere 1:197-201.
Imura, N., E. Sukegawa, S.-K. Pan, K. Nagao, J.-Y. Kim,
  T. Kwan, and T. Ukita (1971) Chemical methylation of
  inorganic mercury with methylcobalamin, a Vitamin B_{12}
  Analog. Science 172:1248-1249.
Iwata, H., H. Okamoto, and Y. Ohsawa (1973) Effect of
  selenium on methylmercury poisoning. Research
  Communications in Chemical Pathology and Pharmacology
  5:673-680.

Jagiello, G. and J.S. Lin (1973) An assessment of the effects of mercury on the meiosis of mouse ova. Mutation Research 17:93-99.

Jarvenpaa, T., M. Tillander, and J.K. Miettinen (1970) Methylmercury: halftime of elimination in flounder, pike and eel. Suomen Kemistilehti B 43:439-442.

Jensen, S. and A. Jernelov (1968) Biological formation of methylmercury in sediments. Nordforsk 14:3-6.

Jensen, S., and A. Jernelov (1969) Biological methylation of mercury in aquatic organisms. Nature 223:753-754.

Jernelov, A. (1970) Release of methylmercury from sediments with layers containing inorganic mercury at different depths. Limnology and Oceanography 15:958-960.

Jernelov, A. (1972) Mercury and food chains. Pages 174-177, Environmental Mercury Contamination, edited by R. Hartung and B.D. Dinman. Ann Arbor, Mich.: Ann Arbor Science Publishers.

Jernelov, A. (1974) Heavy metals, metalloids and synthetic organics. Pages 799-815, The Sea: Ideas and Observations on Program in the Study of the Seas. Volume 5, Marine Chemistry, edited by E.D. Goldberg. New York: John Wiley & Sons.

Jernelov, A., L. Lander, and T. Larsson (1975) Swedish perspectives on mercury pollution. Journal of the Water Pollution Control Federation 47:810-822.

Jernelov, A. and H. Lann (1971) Mercury accumulation in food chains. Oikos 22:403-406.

Jernelov, A., E.L. Lien, and J.M. Wood (1972) Analysis of St. Clair River sediments. Unpublished report, Ontario Water Resources Commission, Toronto, Ontario, Canada.

Jervis, R.E., D. Debrun, W. LePage, and B. Tiefenbach (1970) Mercury Residues in Canadian Foods, Fish, Wildlife. National Health Grant Project No. 605-7-510, Department of Chemical Engineering and Applied Chemistry. Ontario, Canada: University of Toronto.

Johnels, A.G., M. Olsson, and T. Westermark (1968) Esox lucius and some other organisms as indicators of mercury contamination in Swedish lakes and rivers. Bull. Off. Int. Epiz. 69:1439-1452.

Johnels, A.G. and T. Westermark (1969) Mercury contamination of the environment in Sweden. Pages 221-241, Chemical Fallout, Current Research on Persistent Pestidices, edited by M.W. Miller and G.G. Berg. Springfield: Thomas.

Johnels, A.G., T. Westermark, W. Berg, P.I. Persson, and B. Sjostrand (1967) Pike (Esox lucius) and some other aquatic organisms in Sweden as indicators of mercury contamination of the environment. Oikos 18:323-333.

Johnson, D.L. and R.S. Braman (1974) Distribution of atmospheric mercury species near ground. Environmental Science and Technology 8:1003-1009.

Johnson, L.G. and R.L. Morris (1971) Pesticide and mercury levels in migrating duck populations. Bulletin of Environmental Contamination and Toxicology 6:513-515.

Johnson, S.L. and W.G. Pond (1974) Inorganic vs. organic Hg toxicity in growing rats: protection by dietary Se but not Zn. Nutrition Reports International 9:135-147.

Jones, J.R.E. (1940) A further study of the relation between toxicity and solution pressure, with Polycelis nigra as test animal. Journal of Experimental Biology 17:408-415.

Junge, C.E. (1974) Residence time and variability of tropospheric trace gases. Tellus 26:477-488.

Kaku, S., S. Kurata, and S. Yamaguchi (1975) Changes in mercury levels in scalp hair of fish eaters after five years. Japanese Journal of Industrial Health 17:38-39.

Kamps, L.R., R. Carr, and H. Miller (1972) Total mercury-monomethylmercury content of several species of fish. Bulletin of Environmental Contamination and Toxicology 8:273-279.

Kazantzis, G. (1976) Biochemical, physiological and clinical manifestations of exposure to toxic metals. Pages 184-198, Effects and Dose-Response Relationships of Toxic Metals: Proceedings from an international meeting organized by the Subcommittee on the Toxicology of metals of the Permanent Commission and International Association on Occupational Health, Tokyo, Nov. 18-23, 1974, edited by G.F. Nordberg. Amsterdam, N.Y.: Elsevier Scientific Pub. Co.

Keckes, S. and J.K. Miettinen (1970) Mercury as a marine pollutant. In FAO Technical Conference on Marine Pollution and its Effect on Living Resources and Fishing. December 9-18, 1970, International Atomic Energy Agency, Laboratory of Marine Radioactivity, Rome: Food and Agriculture Organization. (Paper No. FIR-MP/70/R.26, accession no. for full conference document 12301-70-WM.)

Keckes, S. and J.K. Miettinen (1972) Mercury as a marine pollutant. Pages 276-289, Marine Pollution and Sea Life, edited by M. Ruivo. London: Fishing News (Books) Ltd.

Khera, K.S. (1973) Reproductive capability of male rats and mice treated with methylmercury. Toxicology and Applied Pharmacology 24:167-177.

Kimura, Y. and V.L. Miller (1964) The degradation of organo-mercury fungicides in soil. Journal of Agricultural and Food Chemistry 12:253-257.

King, J.B. and J.B. Lauckhardt (1973) Mercury in pheasant and other birds from eastern Washington.

Pages 157-166, Mercury in the Western Environment, edited by D.R. Buhler. Corvallis, Oreg.: Continuing Education Publications.

Kirkpatrick, D.C. and D.E. Coffin (1974) The trace metal content of representative Canadian diets in 1970-1971. Canadian Institute of Food Science and Technology Journal 7:56-58.

Kitamura, S. (1968) Determination on mercury content in bodies of inhabitants, cats, fishes, and shells in Minamata District and in the mud of Minamata Bay. Pages 257-266, Minamata Disease, edited by M. Kutsuna. Kumamoto, Japan: Kumamoto University Press.

Klein, D.H., A.W. Andren, J.A. Carter, J.F. Emery, C. Feldman, W. Fulkerson, W.S. Lyon, J.C. Ogle, Y. Taimi, R.I. Van Hook, and N. Bolton (1975) Pathways of thirty-seven trace elements through coal-fired power plant. Environmental Science and Technology 9:973-979.

Knapik, M. (1969) The effect of the $HgNO_3$ content in a water medium upon the survival of certain crustaceans species. Acta Biologica Cracoviensia 12:17-27.

Koeman, J.H., W.H.M. Peeters, C.H.M. Koudstaal-Hol, P.S. Tjioe, and J.J.M. de Goeij (1973) Mercury-selenium correlations in marine mammals. Nature 245:385-386.

Koeman, J.H. and H. van Genderen (1972) Tissue levels in animals and effects caused by chlorinated hydrocarbon insecticides, chlorinated biphenyls and mercury in the marine environment along The Netherlands' coast. Pages 428-435, Marine Pollution and Sea Life, edited by M. Ruivo. London: Fishing New (Books) Ltd.

Koeman, J.H., J.A.J. Vink, and J.J.M. deGoeij (1969) Cause of mortality in birds of prey and owls in the Netherlands in the winter of 1968-1969. Ardea 57:67-73.

Kolbye, A.C., Jr. (1970) Testimony presented at the Hearings before the Subcommittee on Energy, Natural Resources and the Environment of the Committee on Commerce on the Effects of Mercury on Man and the Environment. Pages 30-40, Part 1. Serial 91-72, 91st Congress, 2nd Session.

Koller, L.D. (1975) Methylmercury: Effect on oncogenic and nononcogenic viruses in mice. American Journal of Veterinary Research 36:1501-1504.

Korringa, P. and P. Hagel (1974) In, Proceedings of the International Symposium on the Problems of Contamination of Man and His Environment by Mercury and Cadmium. Luxembourg 3-5 July, 1973. Luxembourg: Commission of European Communities.

Kosta, L., A.R. Byrne, and V. Zelenko (1975) Correlation between selenium and mercury in man following exposure to inorganic mercury. Nature 254:238-239.

Kothny, E.L. (1973) The three-phase equilibrium of mercury in nature. Pages 48-80, Trace Elements in the

Environment, edited by R.T. Gould, Advances in Chemistry Series No. 123. Washington, D.C.: American Chemical Society.

Kreitzer, J.F. (1974) Residues of organochlorine pesticies, mercury, and PCB's in mourning doves from eastern United States, 1970-1971. Pesticides Monitoring Journal 7:195-199.

Kurland, L.T., S.N. Faro, and H. Siedler (1960) Minamata disease. World Neurology 1:370-395.

Laarman, P.W., W.A. Willford, and J.R. Olson (1976) Retention of mercury in the muscle of yellow perch (Perca flavesens) and rock bass (Ambloplites repestrus). Transactions of the American Fisheries Society 105:296-300.

Landner, L. (1971) Biochemical model for the biological methylation of mercury suggested from methylation studies in vivo with Neurospora crassa. Nature 230:452-454.

Langley, D.G. (1971) Mercury methylation in aquatic environment (sic). Paper presented at 162nd national meeting, American Chemical Society, September 12-17, 1971. Division of Water, Air, and Waste Chemistry Paper No. 075 (Abstract only.). Washington, D.C.: American Chemical Society.

Lee, D.F. and J.A. Roughan (1970) Pesticide residues in foodstuffs in Great Britain. XIV. Mercury residues in potatoes. Pesticide Science 1:150-151.

Lee, I.P. and R.L. Dixon (1975) Effects of mercury on spermatogenesis studied by velocity sedimentation cell separation and serial mating. Journal of Pharmacology and Experimental Therapeutics 194:171-181.

Levander, O.A. and L.C. Argrett (1969) Effects of arsenic, mercury, thallium and lead on selenium metabolism in rats. Toxicology and Applied Pharmacology 14:308-314.

Lindberg, S.E. and R.C. Harriss (1973) Mercury-organic matter associations in estuarine sediments and interstitial water. Environmental Science and Technology 8:459-462.

Lindberg, S.E., A.W. Andren, and R.C. Harriss (1975) Geochemistry of mercury in the estuarine environment. Pages 64-107, Estuarine Research. Chemistry, Biology and the Estuarine System, Volume I, edited by E.L. Cronin. New York: Academic Press.

Lockhart, W.L., J.F. Uthe, A.R. Kenney, and P.M. Mehrle (1972) Methylmercury in northern pike (Esox lucius): distribution, elimination, and some biochemical characteristics of contaminated fish. Journal of the Fisheries Research Board of Canada 29:1519-1523.

Lofroth, G. (1973) The mercury problem: a review at midway. Pages 63-70, Trace Substances in Environmental Health - VI, Proceedings of University of Missouri's

6th Annual Conference on Trace Substances in Environmental Health, edited by D.D. Hemphill. Columbia, Mo.: University of Missouri.

MacLeod, J.C. and E. Pessah (1973) Temperature effects on mercury accumulation toxicity and metabolic rate in rainbow trout (Salmo gairdneri). Journal of the Fisheries Research Board of Canada 30:485-492.

Magos, L., A.A. Tuffery, and T.W. Clarkson (1964) Volatilization of mercury by bacteria. British Journal of Industrial Medicine 21:294.

Makhonina, G.I. and E.A. Gileva (1968) Accumulations of Zn-65, Cd-115 and Hg-203 by freshwater plants and the effects of EDTA on the accumulation coefficients for these radioisotopes. Trudy Instituta Ekologii Rastenii i Zhivotnykh 61:72-78. (Chemical Abstracts 70:35161a.)

Marsh, D.O., G. Myers, T.W. Clarkson, L. Amin-Zaki, and S.T. Tikritio (In press) Fetal methylmercury poisoning: new data on clinical and toxicological aspects. (To be published in Transactions of the American Neurological Association.)

Marsh, D.O., M.D. Turner, J.C. Smith, J.W. Choi, and T.W. Clarkson (1974) Methylmercury in human populations eating large quantities of marine fish II. American Samoa. Pages 235-239, Proceedings of the 1st Congreso Internacional del Mercurio, Barcelona, 6-10 de Mayo de 1974. Tomo I. Madrid, Spain: Fabrica Nacional de Moneda y Timbre.

Marsh, D.O., M.D. Turner, and J.C. Smith (1975) Loaves and Fishes: Some Aspects of Methylmercury in Foodstuffs. Submitted to the Food and Drug Administration, U.S. DHEW, under Mercury in Fish, Part 122.200, No. 36, by the University of Rochester School of Medicine and Dentistry. Rochester, N.Y.: The University of Rochester. (Unpublished report available from the USDA, Room 465, 5600 Fishers Lane, Rockville, Md. 20857.)

Martin, W.E. (1972) Mercury and lead residues in starlings. Pesticides Monitoring Journal 6:27-32.

Martin, J.H., P.D. Elliott, V.C. Anderlivi, D. Girvin, S.A. Jacob, R.W. Riseborough, R.L. Delong, and W.G. Gelmartin (1976) Mercury-selenium-bromine imbalances in premature parturient Californian sea lions. Marine Biology 35:91-104.

Martin, W.E. and P.R. Nickerson (1973) Mercury, lead, cadmium, and arsenic residues in starlings - 1971. Pestcides Monitoring Journal 7:67-72.

Massaro, E.J. and F.J. Giblin (1972) Uptake, distribution and concentration of methylmercury by rainbow trout (Salmo gairdneri) tissues. Pages 107-114, Trace Substances in Environmental Health, edited by D.D. Hemphill. Columbia, Mo.: University of Missouri.

Mathew, C. and Zainab Al-Doori (1976) The mutagenic effect of the mercury fungicide Ceresan M in Drosophila melanogaster. Mutation Research 40:31-35.

Matida, Y., H. Kumada, S. Kimura, T. Nose, M. Yokota, and H. Kawatsun (1971) Toxicity of mercury compounds to aquatic organisms and accumulation of the compound by the organisms. Bulletin of the Freshwater Fisheries Research Laboratory of Tokyo 21:197-227.

Matson, R.S., G.E. Mustoe, and S.B. Chang (1972) Mercury inhibition on lipid biosynthesis in freshwater algae. Environmental Science and Technology 6:158-160.

McCarthy, J.H. Jr., W.W. Vaughn, R.E. Learned, and J.L. Meuschke (1969) Mercury in soil gas, and air--A potential tool in mineral exploration. Circular 609, U.S. Geological Survey. Washington, D.C.: U.S. Geological Survey.

McKone, C.E., R.G. Young, C.A. Bache, and D.J. Lisk (1971) Rapid uptake of mercuric ion by goldfish. Environmental Science and Technology 5:1138-1139.

Miettinen, J.K., M. Tillander, K. Rissanen, V. Miettinen, and E. Minkkinen (1968) The excretion by fish, mussel, mollusc and crayfish of methyl mercury nitrate and phenyl mercury nitrate introduced orally or injected into musculature. Paper presented at the Northern Mercury Symposium of Nordforak, Stockholm, October 10-11.

Miettinen, J.K., M. Tillander, K. Rissanen, V. Miettinen, and Y. Ohmoso (1969) Distribution and excretion rate of phenyl- and methylmercury nitrate in fish, mussels, moslluscs and crayfish. Pages 474-478, Proceedings of the 9th Japanese Conference on Radioisotopes. Tokyo: Japan Industrial Forum, Inc.

Miettinen, J.K., M. Heyraud, and S. Keckes (1972b) Mercury as hydrospheric pollutant. II. Biological half-time of methyl mercury in four Mediterranean species: a fish, a crab and two molluscs. Pages 295-298, Marine Pollution and Sea Life, edited by M. Ruivo. London: Fishing News (Books) Ltd.

Miettinen, V., E. Blankenstein, K. Rissanen, M. Tillander, J.K. Miettinen, and M. Valtonen (1972a) Preliminary study on the distribution and effects of two chemical forms of methylmercury in pike and rainbow trout. Pages 298-303, Marine Pollution and Sea Life, edited by M. Ruivo. London: Fishing News (Books) Ltd.

Miller, G.E., P.M. Grant, R. Kishore, F.J. Steinkruger, F.S. Rowland, and V.P. Guinn (1972) Mercury concentrations in museum specimens of tuna and swordfish. Science 175:1121-1122.

Ministry of the Environment (MOE) (1977) The Decline in Mercury Concentration in Fish from Lake St. Clair, 1970-1976. Report No. AQ 577-3, Laboratory Services

Branch, Ministry of the Environment, Rexdale, Ontario, Canada.

Monier-Williams, G.W. (1950) Mercury. Pages 453-468, Trace Elements in Food, 2nd Edition. New York: John Wiley & Sons.

Morrison, A.B. (1971) The Canadian approach to acceptable daily intakes of mercury in foods. Pages 157-164, Proceedings of the Symposium on Mercury in Man's Environment, 15-16 February 1971. Ottawa, Canada: Royal Society of Canada.

Mortimer, D.C. and A. Kudo (1975) Interactions between aquatic plants and bed sediments in mercury uptake from flowing water. Journal of Environmental Quality 4:491-495.

Muto, T. and T. Suzuki (1969) Analytical results of residual mercury in the Japanese storks, Ciconia ciconia boyciana Swinhoe, which died at Obama and Toyooka regions. Japanese Journal of Applied Entomology and Zoology 11:15-20, 1967; Biological Abstracts 50:132151, 1969.

Nakai, S. and I. Machida (1973) Genetic effect of organic mercury on yeast. Mutation Research 21(6):348. (Abstract.)

Nakazawa, N., F. Makino, and S. Okada (1975) Acute effects of mercuric compounds on cultured mammalian cells. Biochemical Pharmacology 24:489-493.

Nash, D.A. (1971) A Survey of Fish Purchases of Socio-economic Characteristics. Annual Report, February 1969 - January 1970. NMFS - Data 62, COM-71-00647, National Marine Fisheries Service, U.S. Department of Commerce. Springfield, Va.: National Technical Information Service.

National Purchase Diary Panel, Inc. (1975) Presentation of the TRF Seafood Consumption Study, September 1973-August 1974. Prepared by the National Purchase Diary for the Tuna Research Foundation. New York: National Purchase Diary, Inc.

National Research Council (1975) Principles for Evaluating Chemicals in the Environment. A report of the Committee for the Working Conference on Principles of Protocols for Evaluating Chemicals in the Environment. Environmental Studies Board, Commission on Natural Resources, and the Committee on Toxicology. Washington, D.C.: National Academy of Sciences.

National Research Council (1977) The Fates of Pollutants: Research and Development Needs. A Report of the Panel on Fates of Pollutants to the Environmental Research Assessment Committee, Environmental Studies Board, Commission on Natural Resources. Washington, D.C.: National Academy of Sciences.

Neujahr, H. and L. Bertilsson (1971) Methylation of mercury compounds by methylcobalamin. Biochemistry 10:2805-2808.

Nishigaki, S. and M. Harada (1975) Methylmercury and selenium in umbilical cords of inhabitants of the Minamata area. Nature 258:324-325.

Nishigaki, S., Y. Tamura, T. Maki, H. Yamada, Y. Shimamura, S. Ochiai, and Y. Kimura (1974) Accumulation of trace elements in fish. I. Relation between the mercury-selenium ration in sea fish muscle and body weight. Toyko Toritsu Eisei Kenkyusho Kenkyu Nempo 25:235-239. (Chemical Abstract 82:153945e, 1975.)

Nordberg, G.F. and P. Strangert (1976) Estimations of a dose-response curve for long-term exposure to methylmercuric compounds in human beings taking into account variability of critical organ concentrations and biological half-time: A preliminary communication. Pages 273-282, Effects and Dose-Response Relationships of Toxic Metals: Proceedings from an international meeting organized by the Sub-committee on Toxicology of the Permanent Commission and International Associations on Occupational Health, Tokyo, 18-23 November, 1974, edited by G.F. Nordberg. Amsterdam, N.Y.: Elsevier Scientific Pub. Co.

Nuzzi, R. (1972) Toxicity of mercury to phytoplankton. Nature 237:38-40.

Ohi, G., S. Nishigaki, H. Seki, Y. Tamura, T. Maki, H. Maeda, S. Ochiai, H. Yamada, Y. Shimamura, and H. Yagyu (1975) Interaction of dietary methylmercury and selenium on accumulation and retention of these substances in rat organs. Toxicology and Applied Pharmacology 32:527-533.

Ohi, G., S. Nishigaki, H. Seki, Y. Tamura, T. Maki, H. Konno, S. Ochiai, H. Yamada, Y. Shimamura, I. Mizoguchi, and H. Yagyu (1976) Efficacy of selenium in tuna and selenite in modifying methylmercury intoxication. Environmental Research 12:49-58.

Olson, K.R. and P.O. Fromm (1973) Mercury uptake and ion distribution in the gills of rainbow trout (Salmo gairdneri): tissue scans with an electron microprobe. Journal of the Fisheries Research Board of Canada 30:1575-1578.

Olson, K.R., H.L. Bergman, and P.O. Fromm (1973) Uptake of methylmercuric chloride and mercuric chloride by trout: A study of uptake pathways into the whole animal and uptake by erythrocytes in vitro. Journal of the Fisheries Research Board of Canada 30:1293-1299.

Olsson, M. (1976) Mercury level as a function of size and age in northern pike, one and five years after the mercury ban in Sweden. Ambio 5:73-76.

Parizek, J. and I. Ostadalova (1967) The protective effect of small amounts of selenite in sublimate intoxication. Experientia 23:142-143.

Parizek, J., I. Benes, I. Ostadalova, A. Babicky, J. Benes, and J. Lener (1969) Metabolic interrelations of trace elements: the effect of some inorganic and organic compounds of selenium on the metabolism of cadmium and mercury in the rat. Physiologia Bohemoslovaca 18:95-103.

Parslow, J.L.F. (1973) Mercury in waders from the wash. Environmental Pollution 5:295-304.

Peyton, T.O., B.E. Suta, and B.R. Holt (1975) Mercury: Human and Ecological Exposure. Prepared for the U.S. Environmental Protection Agency, Contract No. 68-01-2940, Task 019. Menlo Park, Calif.: Stanford Research Institute.

Pierce, P., J.F. Thompson, W.H. Likosky, L.N. Nickey, W.F. Barthel, and A.R. Hinman (1972) Alkyl mercury poisoning in humans, report of an outbreak. Journal of the American Medical Association 220:1439-1442.

Potter, S.D. and G. Matrone (1973) Effect of selenite on the toxicity and retention of dietary methyl mercury and mercuric chloride. Federation Proceedings (Abstr. No. 3997) 32:929.

Potter, S.D. and G. Matrone (1974) Effect of selenite on the toxicity of dietary methylmercury and mercuric chloride in the rat. Journal of Nutrition 104:638-647.

Pyefinch, K.A. and J.C. Mott (1948) The sensitivity of barnacles and their larvae to copper and mercury. Journal of Experimental Biology 25:276-298.

Raeder, M.G. and E. Snekvik (1941) Quecksilberhalt mariner Organismen. Kongelige Horske Videnskabers Selskab Forhandlinger 13:169-172.

Ramel, C. (1967) Genetic effects of organic mercury compounds. Hereditas 57:445-447.

Ramel, C. (1972) Genetic effects. Pages 169-181, Mercury in the Environment; an Epidemiological and Toxicological Appraisal, edited by L. Friberg and J. Vostal. The Chemical Rubber Co. Cleveland. Cleveland, Ohio: CRC Press.

Reinert, R.E., L.J. Stone, and W.A. Willford (1974) Effect of temperature on the accumulation of methylmercury chloride, and pp DDT by rainbow trout (Salmo gairdneri) Journal of the Fisheries Research Board of Canada 31:1649-1652.

Research Committee on Minamata Disease (1975) Pathological, clinical and epidemiological research about Minamata Disease, 10 years after. Kumamoto, Japan: Kumamoto University.

Rivers, J.B., J.E. Pearson, and C.D. Shultz (1972) Total and organic mercury in marine fish. Bulletin of Environmental Contamination and Toxicology 8:257-266.

Rucker, R.R. and D.F. Amend (1969) Absorption and retention of organic mercurials by rainbow trout and chinook and sockeye salmon. Progressive Fish and Culturist 31:197-201.

Ryther, J.H. (1969) Photosynthesis and fish production in the sea. Science 166:72-76.

Saperstein, A.M. (1973) Mercury in benthopelagic fish. Science 180:133.

Schottel, J., A. Mandal, D. Clark, and S. Silver (1974) Volatilization of mercury and organomercurials determined by inducible R-factor systems in enteric bacteria. Nature 251:335-337.

Schrauzer, G., J.A. Seck, R.J. Holland, T.M. Beckham, E.M. Rubin, and J.W. Sibert (1973) Reductive dealkylation of alkylcobaloximes, alkylcobalamins, and related compounds. Simulation of corrin dependent reductase and methyl group transfer reactions. Bio-inorganic Chemistry 2:93-124.

Schrauzer, G., J. Weber, R. Holland, T. Beckham, and R. Ho (1971) Alkyl group transfer from cobalt to mercury: Reaction of alkylcobalamins, alkyl-cobaloximes, and of related compounds with mercuric acetate. Tetrahedron Letters 3:275-277.

Schroeder, H.A. and M. Mitchener (1975) Life-term effects of mercury, methylmercury, and nine other trace metals on mice. Journal of Nutrition 105:452-458.

Sell, J.L. and F.G. Horani (1976) Influence of selenium on toxicity and metabolism of methylmercury in chicks and quail. Nutrition Reports International 14:439-447.

Sergeant, D.E. and F.A.J. Armstrong (1973) Mercury in seals from eastern Canada. Journal of the Fisheries Research Board of Canada 30:843-846.

Shacklette, H.T., J. Boerngen, and R.L. Turner (1971) Mercury in the environment - Surficial materials of the coterminous United States. USGS Circular. Washington, D.C.: U.S. Geological Survey.

Shimp, N.F., H.V. Leland, and W.A. White (1970) Distribution of major, minor, and trace constituents in unconsolidated sediments. Environmental Geology Notes No. 32, Illinois State Geological Survey.

Simpson, R.E., W. Horwitz, and C.A. Roy (1974) Residues in food and feed - surveys of mercury levels in fish and other foods. Pesticides Monitoring Journal 7:127-138.

Skerfving, S. (1974) Methylmercury exposure, mercury levels in blood and hair and health status in Swedes consuming contaminated fish. Toxicology 2:3-23.

Skerfving, S., K. Hansson, C. Mangs, J. Lindsten, and N. Ryman (1974) Methylmercury-induced chromosome damage in man. Environmental Research 7:83-98.

Smart, N.A. (1968) Use and residues of mercury compounds in agriculture. Residue Reviews 23:1-36.

Smart, N.A. and A.R.C. Hill (1968) Pesticide residues in food in Great Britain, VI. Mercury residues in rice. Journal of the Science of Food and Agriculture 19:315-316.

Smith, F.A. (1973) Preliminary studies of mercury tissue levels from game birds and fish in Utah. Pages 140-145, Mercury in the Western Environment, edited by D.R. Buhler. Corvallis, Oreg.: Continuing Education Publications.

Smith, T.G. and F.A.J. Armstrong (1975) Mercury in seals, terrestrial carnivores and principal food items of the Inuit, from Holman NWT. Journal of the Fisheries Research Board of Canada 32:795-801.

Snyder, R.D. (1971) Congenital mercury poisoning. New England Journal of Medicine 18:1014.

Somers, E. (1971) Mercury contamination in foods. Pages 99-106, Proceedings of the Symposium on Mercury in Man's Environment, February 15-16, 1971. Ottawa, Canada: Royal Society of Canada.

Spangler, W.J., J.L. Spigarelli, J.M. Rose, R.S. Flippin, and H.M. Miller (1973a) Degradation of methylmercury by bacteria isolated from environmental samples. Applied Microbiology 25:488-493.

Spangler, W.J., J.L. Spigarelli, J.M. Rose, and H.M. Miller (1973b) Methylmercury: bacterial degradation in lake sediments. Science 180:192-193.

Spyker, J.M., S.B. Sparber, and A.M. Goldberg (1972) Subtle consequences of methylmercury exposure: Behavioral deviations in offspring of treated mothers. Science 177:621-623.

Spyker, J.M. and M. Smithberg (1972) Effects of methylmercury on prenatal development in mice. Teratology 5:181-189.

Stadtman, T.C. (1974) Selenium biochemistry. Science 183:915-922.

Stillings, B.R., H. Lagally, J.H. Soares, and D. Miller (1972) Effect of cystine and selenium on the toxicological effects of methylmercury in rats. Page 206, Proceedings of the IX International Congress of Nutrition. Summaria, Mexico City.

Stillings, B.R., H. Lagally, P. Bauersfeld, and J. Soares (1974) Effect of cystine, selenium, and fish protein on the toxicity and metabolism of methylmercury in rats. Toxicology and Applied Pharmacology 30:243-254.

Stock, A. (1938) Die mikroanalyltische Bestimmung des Quecksilbers und ihre Anwendung auf hygienische und medizinische Fragen. Svensk Kemisk Tidskrift 50:342-350.

Stock, A. and F. Cucuel (1934) Die Verbreitung des Quecksilbers. Naturwissenschaften 22:390-393.

Stoewsand, G.S., C.A. Bache, and D.J. Lisk (1974) Dietary selenium protection of methylmercury intoxication of Japanese quail. Bulletin of Environmental Contamination and Toxicology 11:152-156.

Stoewsand, G.S., J.L. Anderson, W.H. Gutenmann, and D.J. Lisk (1977) Form of dietary selenium on mercury and selenium tissue retention and egg production in Japanese quail. Nutrition Reports International 15:81-87.

Su, M. and G.T. Okita (1976a) Behavioral effects on the progeny of mice treated with methylmercury. Toxicology and Applied Pharmacology 38:195-205.

Su, M. and G.T. Okita (1976b) Embryocidal and teratogenic effects of methylmercury in mice. Toxicology and Applied Pharmacology 38:207-216.

Sugiura, Y., Y. Hojo, Y. Tamai, and H. Tanaka (1976) Selenium protection against mercury toxicity. Binding of methylmercury by the selenohydryl-containing ligand. Journal of the American Chemical Society 98:2339-2341.

Sumino, K., R. Yamamoto, and S. Kitamura (1977) A role of selenium against methylmercury toxicity. Nature 268:73-74.

Summers, A.O. and S. Silver (1972) Mercury resistance in a plasmid-bearing strain of Escherichia coli. Journal of Bacteriology 112:1228-1236.

Summers, A.O. and L.I. Sugarman (1974) Cell-free mercury (II)-reducing activity in plasmid-bearing strain of Escherichia coli. Journal of Bacteriology 119:242-249.

Suter, K.E. (1975) Studies on the dominant-lethal and fertility effects of the heavy metal compounds methylmercuric hydroxide, mercuric chloride, and cadmium chloride in male and female mice. Mutation Research 30:365-374.

Suzuki, T., T. Takemoto, H. Shimano, T. Miyama, H. Katsunuma, and Y. Kagawa (1971) Mercury content in the blood in relation to dietary habit of the woman without any occupational exposure to mercury. Industrial Health Kawasaki 9:1-8.

Takeuchi, T. and K. Eto (1975) Minamata Disease; chronic occurrence from pathological viewpoints. Pages 28-62, Studies on the Health Effects of Alkylmercury in Japan. Japan: Environment Agency.

Takeuchi, T., F.M. D'Itri, P.V. Fischer, C.S. Annett, and M. Okabe (1977) The outbreak of Minamata Disease (methyl mercury poisoning) in cats on northwestern Ontario reserves. Environmental Research 13:215-228.

Tamura, Y., T. Maki, H. Yamada, Y. Shimamura, S. Ochiai, S. Nishigaki, and Y. Kimura (1975) Trace elements in marine fish. III. Accumulation of selenium and mercury

in various tissues of tuna. Tokyo Toritsu Eisei
Kenkyusho Kenkyu Nempo 26:200-204. (Chemical Abstracts
84:162004r, 1976.)

Tanner, J.T., M.H. Friedman, and D.N. Lincoln (1972)
Mercury content of common foods determined by neutron
activation analysis. Science 177:1102-1103.

Task Group on Metal Accumulation (1973) Accumulation of
toxic metals with special reference to their
absorption, excretion and biological half-times.
Environmental Physiology and Biochemistry 3:65-107.

Tezuka, T. and K. Tonomura (1976) Purification and
properties of an enzyme catalyzing the splitting of
carbon-mercury linkages from mercury-resistant
Pseudomonas K-62 strain. Japanese Journal of
Biochemistry 80:79-87.

Thomas, R.L. (1973) The distribution of mercury in the
surficial sediments of Lake Huron. Canadian Journal of
Earth Sciences 10:194-204.

Thomas, R.L., J.M. Jaquet, and A. Murdock (1975)
Sedimentation process and associated changes in
surface sediment trace metal concentrations in Lake
St. Clair, 1970-1974. Pages B97-B103, Proceedings of
the International Conference on Heavy Metals in the
Environment. Toronto, Ontario, Canada.

Tillander, M., J.K. Miettinen, and I. Koivisto (1972)
Excretion rate of methyl mercury in the seal (Pusa
hispida). Pages 303-305, Marine Pollution and Sea
Life, edited by M. Ruivo. London: Fishing News (Books)
Ltd.

Tsubaki, T. (1971) Clinical and epidemiological aspects
of organic mercury intoxication. Pages 131-136,
Proceedings of the Symposium on Mercury in Man's
Environment, 15-16 February 1971. Ottawa, Canada: The
Royal Society of Canada.

Tsubaki, T. and K. Irukayama, eds. (1977) Minamata
Disease (Methylmercury poisoning in Minimata and
Niigata, Japan). New York: Elsevier Scientific
Publishing Company, Kodansha, Ltd.

Turner, M.D., D.O. Marsh, C.E. Rubio, J. Chiriboga, C.C.
Chiriboga, J.C. Smith, and T.W. Clarkson (1974)
Methylmercury in populations eating large quantities
of marine fish I. Northern Peru. Pages 229-234, The
Proceedings of the 1st Congreso Internacional del
Mercurio, Barcelona, 6-10 de Mayo de 1974, Tomo I.
Madrid, Spain: Fabrica National de Moneda y Timbre.

Ueda, K., S. Yamanaka, M. Kawai, and K. Tohjo (1975a)
Effects of selenium on methylmercury poisoning in
rats. Igaku To Seibutsugaku 90:15-20 (Chemical
Abstracts 83:91887x.)

Ueda, K., M. Kawai, S. Yamanaka, K. Tohjo, and N. Someya
(1975b) Paper presented at the 18th International

Congress on Occupational Health, Brighton, September 14-19.

Ui, J. and S. Kitamura (1971) Mercury in the Adriatic. Marine Pollution Bulletin 2:56-58.

U.K. Department of the Environment (1976) Environmental Mercury and Man. A Report of an Inter-Departmental Working Group on Heavy Metals. Department of the Environment, Central Pollution Paper No. 10. Unit on Environmental Pollution. London: Her Majesty's Stationary Office.

U.K. Ministry of Agriculture, Fisheries and Food (1971) Survey of mercury in food. Working Party on the Monitoring of Foodstuffs for Mercury and Other Heavy Metals, First Report. London: Her Majesty's Stationary Office.

U.K. Ministry of Agriculture, Fisheries and Food (1973) Survey of mercury in food: A supplementary report. Working Party on the Monitoring of Foodstuffs for Mercury and Other Heavy Metals. London: Her Majesty's Stationary Office.

Underdal, B. (1969) Study of mercury in some food stuffs. Nordisk Hygienisk Tidskrift 50:60-63.

Underdal, B. and T. Hastein (1971) Mercury in fish and water from a river and a fjord in the Kragero region, south Norway. Oikos 22:101-109.

Unlu, M.Y., M. Heyraud, and S. Keckes (1972) Mercury as a hydrospheric pollutant. I. Accumulation and excretion of 203 $HgCl_2$ in Tapes decussatus L. Pages 292-295, Marine Pollution and Sea Life, edited by M. Ruivo. London: Fishing News (Books) Ltd.

U.S. Environmental Protection Agency (1973) Water Quality Criteria 1972 Ecological Research Series. A report of the Committee on Water Quality Criteria, Environmental Studies Board, National Academy of Sciences. EPA/R3/73/033. Washington, D.C.: U.S. Government Printing Office.

U.S. Environmental Protection Agency (1975a) Chemical Analysis of Interstate Carrier Water Supply Systems. EPA-430/9-75-005. Washington, D.C.: U.S. Environmental Protection Agency.

U.S. Environmental Protection Agency (1975b) Materials Balance and Technology Assessment of Mercury and its Compounds on National and Regional Bases. October 1975, Final Report. Prepared for Environmental Protection Agency, Office of Toxic Substances, by URS Research Company. EPA-560/3-75-007. Washington, D.C.: U.S. Environmental Protection Agency.

Uthe, J.F., F.M. Atton, and L.M. Roger (1973) Uptake of mercury by caged rainbow trout (Salmo gairdneri) in the south Saskatchewan river. Journal of the Fisheries Research Board of Canada 30:643-650.

Vermeer, K. (1971) A survey of mercury residues in aquatic bird eggs in the Canadian prairie provinces. Transactions of the North American Wildlife Conference 36:138-150.

Vermeer, K. and F.A.J. Armstrong (1972) Mercury in Canadian prairie ducks. Journal of Wildlife Management 38:179-182.

Vermeer, K., F.A.J. Armstrong, and D.R.M. Hatch (1973) Mercury in aquatic birds at Clay Lake, western Ontario. Journal of Wildlife Management 37:58-61.

Vernet, J.P. and R.L. Thomas (1972) The occurrence and distribution of mercury in the sediments of the Petit-Lac (western Lake Geneva). Ecologae Geologicae Helvetiae 65:307-316.

Vonk, J.W. and A.K. Sijpesteijn (1973) Methylation of mercuric chloride by pure cultures of bacteria and fungi. Antonie van Leenuwenhoek Journal of Microbiology and Serology 39:505-513.

Wahlberg, P., E. Karppanen, K. Henriksson, and D. Nyman (1971) Human exposure to mercury from goosander eggs containing methylmercury. Acta Medica Scandinavica 189:235-239.

Wanntorp, H., K. Borg, E. Hanko, and K. Erne (1967) Mercury residues in wood-pigeons (Columba p. palumbus L.) in 1964 and 1966. Nordisk Veterinalmedicia 19:474-477.

Weigand, J.P. (1973) Mercury in Hungarian partridge and in their northcentral Montana environment. Pages 172-185, Mercury in the Western Environment, edited by D.R. Buhler. Corvallis, Oreg.: Continuing Education Publications.

Weiss, H.V., M. Koide, and E.D. Goldberg (1971) Mercury in a Greenland ice sheet: evidence of recent input by man. Science 174:692-694.

Weiss, H.V., S. Yamamoto, T.E. Crozier, and J.H. Matthewson (1972) Mercury: Vertical distribution at two locations in the eastern tropical Pacific Ocean. Environmental Science and Technology 6:644-645.

Weiss, H.J., K. Bertine, M. Koide, and E.D. Goldberg (1975) The chemical composition of a Greenland glacier. Geochimica et Cosmochimica Acta 39:1-10.

Welsh, S.O., J.H. Soares, B.R. Stilling, and H. Lagally (1973) Effects of mercury and selenium on serum transaminase levels of quail, hens and rats. Nutrition Reports International 8:419-429.

Welsh, S.O. and J.H. Soares (1974) The bioavailability of mercury in the tissue of hens fed methylmercuric chloride. Federal Proceedings (Abstr. No. 2545) 33:660.

Welsh, S.O. and J.H. Soares (1976) The protective effect of vitamin E and selenium against methylmercury

toxicity in Japanese quail. Nutrition Reports
International 13:43-51.

Welsh, S. (1974) Physiological Effects of Methylmercury
Toxicity: Interaction of Methylmercury with Selenium,
Tellurium and Vitamin E. Ph.D. Thesis. College Park,
Md.: University of Maryland.

Westermark, T. (1965) Mercury in aquatic organisms.
Pages 25-76, The Mercury Problem in Sweden. Stockholm:
Royal Swedish Ministry of Agriculture.

Westoo, G. (1965) Mercury in foodstuffs - is there a
great risk of poisoning? Var Foeda 4:1-6.

Westoo, G. (1968) Determination of methylmercury salts
in various kinds of biological materials. Acta Chemica
Scandinavica 22:2277-2280.

Westoo, G. (1969a) Mercury and methylmercury levels in
some animal products. August 1967-October 1969. Var
Foeda 21:137-154.

Westoo, G. (1969b) Methylmercury compounds in animal
foods. Pages 75-93, Chemical Fallout. Current Research
on Persistent Pesticides, edited by M.W. Miller and
G.G. Berg. Springfield: Thomas.

Westoo, G. (1973) Methylmercury as a percentage of total
mercury in flesh and viscera of salmon and sea trout
of various ages. Science 181:567-568.

Westoo, G. (1975) Nashville Conference. Methylmercury
Analysis (K. Sumino). Discussion. Pages 47-50, Heavy
Metals in the Aquatic Environment, edited by P.A.
Krenkel. Oxford: Pergamon Press.

Wetzel, R.G. (1975) Limnology. Philadelphia: W.B.
Saunders Company.

Wiemeyer, S.N., B.M. Mulhern, F.J. Ligas, R.J. Hensel,
J.E. Mathisen, F.C. Robards, and S. Postupalsky (1972)
Residues or organochlorine pesticides, polychlorinated
biphenyls, and mercury in bald eagle eggs and changes
in shell thickness - 1969 and 1970. Pesticides
Monitoring Journal 6:50-55.

Williston, S.H. (1968) Mercury in the atmosphere.
Journal of Geophysical Research 73:7051-7055.

Wisely, D. and R.A.P. Blick (1967) Mortality of marine
invertebrates larvae in mercury, copper and zinc
solutions. Australian Journal of Marine and Freshwater
Research 18:63-72.

Wobeser, G., N.O. Nielsen, R.H. Dunlop, and F.M. Atton
(1970) Mercury concentrations in tissues of fish from
the Saskatchewan River. Journal of the Fisheries
Research Board of Canada 27:830-838.

Wollast, R., G. Billen, and F.T. Mackenzie (1975)
Behaviour of mercury in natural systems and its global
cycle. Pages 145-166, Ecological Toxicology Research:
Effects of Heavy Metal and Organohalogen Compounds.
Proceedings of a NATO Science Committee Conference,

edited by A.D. McIntyre and C.F. Mills. New York & London: Plenum Press.

Wood, J.M. (1971) Environmental Pollution by Mercury. Pages 39-56, Advances in Environmental Science and Technology, Volume II, edited by J.M. Pitts and R.L. Metcalf. New York: Wiley Interscience.

Wood, J.M. (1973) Toxic elements in the environment. Pages 1-8, Revue Internationale D'Oceanographie Medicale, edited by M. Aubert. Centre d'Etudes et de Recherches de Biologic et d'Oceanographic Materiale 31(32):1-8.

Wood, J.M. (1974) Biological cycles for toxic elements in the environment. Science 183:1049-1052.

Wood, J.M. (1975a) Biological cycles for elements in the environment. Die Naturwissenschaften 62(8):357-364.

Wood, J.M. (1975b) Metabolic cycles for toxic elements, in the environment: a study of kinetics and mechanism. Pages 105-112, Heavy Metals in the Aquatic Environment, edited by P.A. Krenkel. Oxford: Pergamon Press.

Wood, J.M. (1976) Les metaux toxiques dans l'environment. La Recherche 7:711-719.

Wood, J.M., F.S. Kennedy, and C.G. Rosen (1968) Synthesis of methyl-mercury compounds by extracts of a methanogenic bacterium. Nature 220:173-175.

World Health Organization (1971) International standards for drinking water, 3rd ed. Geneva: World Health Organization.

World Health Organization (1972) Evaluation of mercury, lead, cadmium, and the food additives amaranth, diethylpyrocarbonate, and octyl gallate. Joint FAO/WHO Expert Committee on Food Additives. Geneva, April 4-12, 1972, WHO Food Additives Series, No. 4. Geneva: World Health Organization.

World Health Organization (1976a) Conference on Intoxication due to Alkylmercury-treated Seed. Baghad, Iraq, 9-13 September 1974. Supplement to Volume 53. Geneva: World Health Organization.

World Health Organization (1976b) Environmental Health Criteria 1, Mercury. Report from a meeting held 4-10 February, 1975. Geneva: World Health Organization.

Yamada, M. and K. Tonomura (1972) Formation of methylmercury compounds from inorganic mercury by Clostridium cochlearium. Journal Fermentation Technology 50:159-166.

Yamaguchi, S., H. Matsumoto, S. Matsuo, S. Kaku, and M. Hoshide (1971) Relationship between mercury content of hair and amount of fish consumed. HSMHA Health Reports 86:904-909.

Yatscoff, R.W. and J.E. Cummins (1975) DNA breakage caused by dimethyl mercury and its repair in a slime mould, Physarum polycephalum. Nature 257:422-423.

Zook, E.G., J.J. Powell, B.M. Hackley, J.A. Emerson, J.R. Brooker, and G.M. Knobl, Jr. (1976) National Marine Fisheries Service preliminary survey of selected seafoods for mercury, lead, cadmium, chromium, and arsenic content. Journal of Agricultural and Food Chemistry 24:47-53.

APPENDIX A

MERCURY ANALYTICAL METHODS[1]

CONSIDERATIONS IN ANALYSIS

Analysis is implicit in every document or article citing mercury concentration data, regardless of whether the source of mercury is experimental, environmental, animal, or human. In the event of contradictory data, it may be important to judge the reliability of the trace level analysis. Hume (1973) has examined the present state of the art of trace metal analysis and its implications; and this discussion follows his general comments.

Hume (1973) points out the general lack of interlaboratory agreement in analyses of blind samples. An example is given from an unpublished report dated 1970 of 13 laboratories that participated in analyzing the trace metal content (not mercury) of three subsamples, each taken from two large, primary samples of seawater. Conventional atomic absorption spectroscopy, neutron activation analysis, or colorimetric procedures were used. The reported values ranged from 3.7 to 47 μg/kg (ppb), with a range of relative standard deviations per set of analyses between 3 and 70 percent. Such results do not indicate that water chemists are particularly inept. Rather, Hume (1973) hastens to mention that the same lack of consistency is, at least to some degree, characteristic of all experimental measurements. However, the discrepancies are not obvious except in demanding procedures such as trace metal analyses where the "signal to noise" ratio is unfavorable and the pitfalls are numerous. Nonetheless, some sources of discrepancies can be isolated, examined, and possibly eliminated.

Sources of Discrepancies in Analytical Results

The operator and instrument variabilities are understandable sources of disagreement and can be controlled somewhat beyond automated procedures and instrumental techniques as are now largely implemented at government agencies like the U.S. Environmental Protection Agency (U.S. EPA 1974). Other sources require an appreciation of the minute quantities of metal measured (Table A.1). A natural mercury concentration of 1 ng/g, or 1 ppb, usually corresponds to less than one atom of mercury for every 10^9 (billion) atoms of other substances. Detection of this tiny but commonly encountered concentration is a result of a multistage process, each step of which provides a means for the introduction of errors. Before the actual instrumental measurement, the sample is usually collected, stored, pretreated for quantitative subsampling, then chemically treated for separation or concentration. A typical chemical treatment for a biological sample is as follows (Smith and Windom 1972): (1) weigh 0.5 g of frozen tissue into a 100-ml beaker, (2) add 4.0 ml of concentrated sulfuric acid and 2.5 ml of concentrated nitric acid, (3) cover beaker with a watch glass and place in a water bath at 58°C overnight, (4) carefully transfer the sample solution to a BOD bottle, washing the original sample beaker with redistilled water, (5) dilute sample solution to 100 ml and add 1 ml of potassium permanganate solution, (6) shake and add additional portions of potassium permanganate solution until the purple color persists at least 15 min, (8) add 2 ml of potassium persulfate solution in 2-ml increments until a clear solution is observed, (9) add 5 ml of stannous sulfate and immediately attach to aeration assembly, and (10) from recorder peak height, find the μg of mercury from a calibration curve.

At each step, metal can be lost or gained. Reagents are notorious for possessing background metal ion concentrations, often far in excess of the sample (U.S. EPA 1972). Possible loss of mercury compounds by volatilization during chemical processing has long been recognized as another potential source of error. Likewise, during storage, mercury can be leached into solution from the laboratory or container materials or be lost to container walls by sorption (Campbell et al. 1972). Thus, one reason for promotion of neutron activation analysis is the accessibility of a blank container count before sample collection (Thatcher and Johnson 1971). A recent publication (Weiss and Chew 1973) presents data indicating a 30 percent loss of induced activity (12 percent to the walls of the vial) in irradiation of unacidified aqueous mercury solutions. The loss is thought to occur by two routes: adsorption and volatilization during irradiation.

TABLE A.1   Units of Weight and Concentration[a]

Description and Unit	Equivalent
Weight:	
1 kilogram (kg)	1,000 grams (g)
1 milligram (mg)	10^{-3} g
1 microgram (μg)	10^{-6} g
1 nanogram (ng)	10^{-9} g
1 picogram (pg)	10^{-12} g
Concentration:	
Weight: weight basis (for foods, body organs, and other solids)	
1 part per million (ppm)	1 mg/kg or 1 μg/g
1 part per billion (ppb)	1 μg/kg or 1 ng/g
1 part per trillion (ppt)	1 ng/kg or 1 pg/g
Weight: volume basis (for water)	
1 mg/liter	approximately 1 μg/g on weight: weight basis
1 μg/liter	approximately 1 ng/g on weight: weight basis
1 ng/liter	approximately 1 pg/g on weight: weight basis
Weight: volume basis (for air)	
1 mg/cubic meter	10^{3} μg/cubic meter or 10^{6} ng/cubic meter

[a]The conversion of a weight:volume relationship to a volume:volume basis depends on the molecular weight of the dispersed substance (200.6 g/mole for elemental mercury). At 25°C and 760 mm Hg pressure, the conversion formula is: ppm by volume = mg/cubic meter × 24.45/molecular weight where mg/cubic meter is the measured concentration of mercury.

In addition to the trace metal, environmental samples contain an unpredictable assortment of inorganic and organic substances that can contribute to analytical interference. Natural chelating agents derived from decomposition of plant and animal matter can prevent water-organic liquid extraction of the metal. Algal or bacterial action on mercury may continue in stored natural fluids while mercury in tissue samples continues to change during decomposition. Both processes can be controlled by proper treatment at the time of collection.

In conclusion, trace mercury analysis in the ng/g range is a demanding procedure, successfully accomplished only by considering and avoiding or compensating for all the known pitfalls. Only analyses that use proven or standardized procedures and report standard deviations as well as calibrations with a known standard or spiked sample should be accepted. Trace metal analysis is presently an extremely active area of research directed at establishing procedures leading to increased sensitivity and accuracy.

Evaluation of Early Analytical Results

Techniques for trace metal analyses have undergone considerable improvement in the past 10 yr, leading to a need for assessment of earlier analyses. The change itself is not as important to recognize as the direction of change. Before the mid-1960s, for instance, many analyses of biological tissue determined total mercury without distinguishing between inorganic and organic forms. In identifying methylmercury as the prevalent form in fish, Westoo (1966) focused attention on the importance of analysis for organomercury compounds. Today, analyses of environmental or biological samples for mercury are more likely to involve detection of the form as well as the quantity. Likewise, as improvements in analytical techniques have been instituted, greater sensitivity, selectivity, and reliability have been attained. These factors lead to improved detection of the mercury in a sample even as the concentrations diminish.

Since more pathways for error lead to apparent low rather than apparent high determinations, it is probably safe to assume that, if earlier analyses were in error, they were more likely to have reported concentrations lower than the true value. Mercury was reported absent because of poor instrument detectability. Also, volatile mercury compounds were lost in the hot digestion of samples, or mercury was not totally recovered because of inefficient oxidizing reagents.

Processing for total mercury analyses often neglected
organic mercury compounds entirely.

On the other hand, the commonly cited mechanism for
higher-than-true mercury determinations, contamination
of reagents and laboratory ware, can contribute parts
per billion, or at most parts per million, in
particularly aggravated cases. Blank determinations
along with purification of reagents and treatment of
labware help eliminate this source of error. Another
important route that leads to apparent high values
involves the loss of mercury from synthetic standard
solutions. They are almost invariably prepared with
distilled or deionized water. As noted in Table A.2,
unless the sample is carefully treated, the half-life of
mercury in distilled water is as little as half that of
mercury in natural water; the loss involves adsorption
onto or diffusion into container walls. Natural water
contains an abundance of soluble ions thought to prolong
the half-life of mercury (Rosain and Wai 1973). Feldman
(1974) noted an apparent increase with time in the
mercury concentration in natural water samples. The
increase was attributable to a decrease with time in the
strength of his synthetic standard solution. His
recommendations for the preservation of standard
solutions are discussed later in the section on chemical
treatment of samples.

A fair portion of the systematic errors in mercury
analysis are now likely to have been detected and
documented. Although there is always a lag between
documentation and implementation, the improved agreement
in interlaboratory analyses (discussed later in this
paper) may be indicative of improved analyses in
general. Likewise, the use of standard methods and
environmental samples lends added confidence to recent
analyses.

ANALYTICAL PROCEDURES

Storage and Preservation

A sample that cannot be analyzed immediately must be
stored. Contrary to expectation, this step is not a
static phase of the analytical procedure; in fact, it
presents another route for the loss of mercury. In this
case "loss" indicates mercury originally present that
is undetectable. Rosain and Wai (1973) studied the rate
of mercury loss from solutions of natural and distilled
water spiked with 25 $\mu g/l$ mercury ion when stored in
stoppered polyethylene, polyvinyl chloride, and soft
glass containers. Severe losses observed at pH 2 and 7
were considerably curtailed by acidification with nitric

TABLE A.2   Half-Life in Days of 26 μg/1 Mercury in Aqueous Solution

Container	pH 7		pH 2
	Distilled Water	Creek Water	Creek Water
Polyethylene	1.58	3.95	4.0
Polyvinyl chloride	1.98	4.38	1.7
Soft glass	2.98	4.44	3.5

SOURCE: Rosain and Wai (1973).

acid to pH 0.5. The half-lives of stored mercury are given in Table A.2. The loss is exponential, with only half the original quantity remaining after as little as 1.5 days. At pH 0.5, the loss was not detected after 4.4 days and was barely detected (2 percent loss) after 15 days. Feldman (1974), working at the 0.1 to 10.0 μg/l range, found losses of mercury from both glass and polyethylene containers in spite of acidification and preservation with potassium permanganate. Addition of 0.01 percent dichromate ion in place of permanganate stabilized the acidified solutions for as long as five months.

Solid biological samples are adequately stored without loss of mercury by freezing immediately after collection (Kopp et al. 1972). Preservation and storage of urine samples have been discussed by Trujillo et al. (1974). Potassium persulfate, added at the time of collection, preserves the sample for several days.

Concentration and Separation

In principle, any mercury concentration, no matter how small, can be determined if a sufficiently large sample is collected and the total mercury separated into a small volume. However, any process, no matter how judiciously applied, represents a route for the loss of mercury.

Evaporation

Concentration and separation are essentially the same process. The first removes the matrix from the metal; the second removes metal from the matrix. The simplest concentration technique is evaporation, a valuable but frequently undesirable method, considering the volatility of many mercury compounds, particularly organomercurials. Nonetheless, it has been used advantageously to concentrate mercurials in benzene solution by evaporation of a portion of the benzene after liquid-liquid extraction of organomercury salts preparatory to gas chromatographic injection (Longbottom et al. 1973).

Solvent Extraction

This technique is widely used in aqueous analyses for separating and concentrating organic mercury compounds and depends upon the differential solubility of mercury compounds in mutually immiscible liquids, for example, benzene and water. Specifically, benzene or toluene is added to a mercury-containing aqueous solution, and the mixture is agitated sufficiently to allow optimum transfer of the solute between the two phases. Then the mixture is allowed to separate, and the organic phase containing mercury is drawn off. This procedure may be repeated with fresh solvent as often as necessary to effect as complete a recovery as possible. Inorganic mercury may be extracted into an organic solvent by adding an organic chelating agent to form an organic mercury complex. Dithizone, widely used to extract inorganic mercury, not only complexes quantitatively but it forms a colored complex that is detected spectrophotometrically (Sandell 1959). The principles and applications of solvent extraction to water analysis are discussed by Andelman (1971).

Amalgamation

Like solvent extraction, amalgamation depends on the differential solubility of a solute, in this case elemental mercury, in each of two different phases. Traces of metallic mercury are immediately soluble in gold and silver. Grids of gold or silver are placed in an airstream to collect airborne elemental mercury; a gold-coated, fritted glass disk has been used to capture mercury by passing air through it at 1 l/min. Similarly, mercury has been collected from acidified water samples by amalgamation on a silver wire (Fishman 1970) or by electrodeposition onto a copper cathode (Doherty and Dorsett 1971), although the presence of sulfide ion in the water causes low results (Fishman 1970). Amalgamated mercury is released (usually into an atomic absorption cell) by heating to drive off the volatile metal.

Carbon Adsorption

Atmospheric mercury can be concentrated on activated carbon. Moffitt and Kupel (1970) used a commercially available impregnated charcoal to trap mercury in industrial atmospheres. Air is pumped through a sampling tube containing two tandem charcoal sections separated from each other and from the atmosphere by glass wood plugs. The glass wood plug at the inlet end of the tube is analyzed for particulate-bound mercury, and adsorbed volatile mercury is determined by atomic adsorption spectroscopy. The effectiveness of the charcoal filter is determined by the absence of mercury on the second section.

Ion Exchange

Ion exchange separation is a well-established analytical technique, although applications to mercury analysis have been few. Selectivity occurs as a result of the relative affinity of a given ion exchange resin for particular ions (Andelman 1971). An anion exchange, resin-loaded filter paper has been used to remove mercury from natural water containing from 0.03 to 6.5 μg/l mercury (Becknell et al. 1971). Both the inorganic and organic forms of mercury are converted to $HgCl_4^{2-}$ before filtering. The mercury-loaded papers are air dried and then sealed in mylar film for irradiation before neutron activation analysis. In another study, mercury collected directly from seawater on a chelating resin was so strongly retained that no reagent was able to elute it completely for analysis by atomic absorption spectroscopy (Riley and Taylor 1968).

Chemical Treatment of Samples

Aqueous Mercury

Mercury in aquatic samples appears as inorganic or organic species either dissolved, sorbed onto particulate matter, or entrained within a particulate matrix. Total mercury determination measures all of these. Thus, samples require pretreatment to release mercury from all the forms in which it is collected and to transform it into a state that is compatible with the measuring technique. Kopp et al. (1972) have described some of the obstacles inherent in the pretreatment procedure for lake and river waters and have also proposed methods for overcoming them. The addition of acids to aqueous samples, for example, generates heat, which can lead to losses of mercury compounds by volatilization. Aquatic materials also require an oxidizing agent strong enough to decompose the organomercury compounds completely. Kopp et al. (1972) demonstrated (1) the ineffectiveness of commonly used potassium permanganate only to release all organically bound mercury, and (2) the success of added potassium persulfate in conjunction with potassium permanganate in total recovery of mercury from solutions containing dissolved phenyl and methylmercuric salts. The procedures now incorporated into the standard methodology for water and wastes (U.S. EPA 1974) are outlined briefly: (1) an aliquot is acidified with sulfuric and nitric acids, (2) potassium permanganate solution is added until the purple color persists for at least 15 min, (3) potassium persulfate is added, and

heat is applied for 2 h at 95°C, (4) sodium chloride-
hydroxylamine sulfate solution is added to reduce excess
permanganate, and (5) stannous sulfate solution is added
just before aeration into the atomic absorption cell.

Biological Mercury

   To analyze for mercury in biological tissue, it is
important to determine the total mercury content and
also the fraction of organomercurial compounds,
methylmercury in particular. Uthe (1971) describes a
thoroughly tested total mercury procedure essentially
like that outlined in the section above of sources of
discrepancy of analytical results except for semi-
automation of the method. The procedure consists of
cold, wet, acid oxidation to decompose the organic
matter and release mercuric mercury, which is then
reduced to elemental mercury, the form aerated from
solution for atomic absorption measurement.
   Alkylmercury determinations require a completely
different approach to preserve the organic-bond formed
during extraction from the matrix. Most techniques are
variations of the original Westoo method (Westoo 1966,
1967, 1968), and they consist basically of organic
solvent extraction and re-extraction. Chemical
treatment precedes extraction to remove interfering
substances such as sulfides and to convert the
organomercury compound to the halide salt (chloride or
bromide), the form most accessible to electron capture
detection after chromatographic separation.
Dialkylmercury compounds such as dimethylmercury are not
as strongly bound to the matrix and are easily extracted
into an organic medium (benzene or toluene) as a first
step before chemical treatment of the remaining sample.
Care is required, however, to prevent loss of the highly
volatile dimethylmercury. This extracted mercury is
then subjected to halide treatment: the
monomethylmercury halide thus produced can then be
analyzed by gas chromatography (Mushak 1973).
   The U.S. Environmental Protection Agency has proposed
for adoption a procedure for methylmercury determination
in biological media as described by Longbottom et al.
(1973). Preservation, storage, chemical treatment,
extraction, and cleanup are included. Copper sulfate is
introduced at the time of collection to preserve the
integrity of mercury compounds until analysis. Fish and
sediment samples are frozen upon collection, thawed
before use, then treated with copper sulfate to displace
methylmercury from its strong, natural inorganic and
organic sulfur bonds. The free methylmercury is
converted to the bromide salt by addition of excess

potassium bromide to a solution strongly acidified with sulfuric acid (pH below 0.5); then it is extracted into a toluene layer. The extract cannot be injected into the gas chromatograph until sulfur compounds are completely removed. This is the purpose of the cleanup step, accomplished by high-efficiency extraction of methylmercury into an aqueous thiosulfate phase, addition of potassium iodide, then re-extraction into benzene or toluene for injection into the gas chromatograph. Any of the extraction or cleanup steps may be repeated for improved recovery.

Airborne Mercury

Mercury appears in air in several forms: as elemental free mercury atoms, as inorganic or organic mercury molecules, and as any of these adsorbed on particulate matter in the air. All forms present a hazard that necessitates their detection and monitoring.

Elemental mercury vapor, the form usually monitored, requires no pretreatment and is detected directly by drawing air through a high sensitivity atomic absorption spectrometer (Jepsen 1973). Alternatively, mercury vapor can be collected before analysis (1) by metering airflow through flasks containing potassium permanganate-sulfuric acid or iodine-potassium iodide solution, (2) by adsorption onto impregnated charcoal, or (3) by deposition on filter materials (Lindstedt and Skerfving 1972, Bogen 1973). Airborne mercury trapped in solution is recovered for analysis by any of the standard pretreatment methods, whereas that deposited on filters, including particulate matter, may be determined directly by neutron activation analysis (Bogen 1973).

These collection systems are inadequate to detect organic mercurials. Linch et al. (1968) found less than 20 percent of dimethylmercury and less than 50 percent of diethylmercury in air retained by an aqueous iodine-potassium iodide reagent, and less than 5 percent of dimethylmercury is retained by potassium permanganate-sulfuric acid solution. Recoveries averaging 98 percent for dimethylmercury in the 2 to 60 μg range, however, were attained with an iodine monochloride reagent. Near quantitative collection of diethylmercury, methylmercuric chloride, and ethylmercuric chloride also was established. Iodine monochloride is considered to be the only known absorbent that is effective for collecting both inorganic and organic mercury (Anonymous 1973a). The final mercury determination, however, requires a method free of interference from iodine. In other words, atomic absorption spectroscopy cannot be used.

Methods of Analysis

The reliability of any analytical procedure can be no better than its least reliable step. Even assuming no systematic errors, the uncertainty of an analysis is the sum of the uncertainties contributed by the individual analytical processes. This uncertainty arises from random error associated with any experimental procedure. For this reason, an analysis is a continual checking process, first to eliminate any systematic error, and second to minimize random error.

The crucial instrumental operation often is the limiting factor in an analysis, and most of the preliminaries are preparatory to the instrumental step. In fact, an analytical method is named for the instrument step even though it oftens requires the least time and effort. Table A.3 shows the common instrumental methods, including a very brief description of sample preparations and measurement techniques.

The sensitivity (column 6) is largely instrument dependent, experimentally determined, and variously defined by different authors. Essentially, the threshold of sensitivity is the smallest quantity of substance that can be determined reliably by the instrumental technique. In atomic absorption spectroscopy, for instance, the sensitivity has been defined as that concentration of analyte that yields 1 percent of a full-scale reading, or that produces an absorption of 1 percent (Minear and Murray 1974), or that is 3 times the standard deviation of blank readings (Anonymous 1974).

The precision of the method is given in column 7 as the relative standard deviation (that is, the standard deviation of a set of sample measurements divided by the concentration of the sample). The sample concentration is also shown since the precision is strongly dependent on it.

Column 8, accuracy, is measured by a procedural calibration technique. Kaiser (1973) has discussed the three main calibration methods for trace determination: with synthetic standard samples, with analyzed standard samples, or by differential additions. Ideally, synthetic standards are prepared and processed with the sample. With mercury, however, it is exceedingly difficult, if not impossible, to prepare synthetic trace samples from truly pure substances and to transfer them into a state that duplicates the natural analytical sample. Use of pre-analyzed natural samples solves this problem but presupposes that the problem of analyzing a natural trace sample with adequate calibration procedures has already been solved (Alvarez 1974). Nonetheless, the National Bureau of Standards (NBS)

biological pre-analyzed samples are available, and their
use is increasing (Anonymous 1973b). The most prevalent
calibration method, by differential additions, is the
one listed here. Small, known quantities of a mercury
compound are added to the unknown samples before
processing. Recovery of this quantity measures the
accuracy of the procedure. The main problem is to
insure that the spike is in the same physical and
chemical condition as the natural mercury component.

Other than the columns discussed, Table A.3 is
intended to be self-explanatory. However, it is not
intended to be a complete description of analytical
methods. For instance, although total mercury is
determined in atomic absorption spectroscopy, this does
not preclude determination of organomercury
concentrations. First, an intact sample is processed;
then a sample in which the organomercurials have been
separated by organic liquid extraction is processed.
The difference between the two determinations gives the
concentration of organomercury compounds. However, the
species of the organomercurial cannot be determined
directly by atomic absorption. A brief description is
given in the following sections of the listed
instrumental methods, preceded by a few well-known
chemical methods. Some newer exploratory techniques are
discussed in the later section on newer methods of
analysis.

Gravimetric Analysis

All gravimetric procedures are essentially alike. The
sample is digested with a reducing solution to convert
all mercury to the elemental form. The mercury is
driven off as a vapor by heating and is collected by
amalgamation on a metal foil or screen. Gold or silver
is usually used, and the mercury is determined by the
difference in weight before and after the amalgamation.
With routine analytical balances, a few mg of mercury
can be determined with reasonable accuracy; with the
newer microbalances, this level can be extended to a few
μg. A variation of this method involves the deposition
of the mercury on a platinum cathode. This is the
method recommended by the American Society for Testing
and Materials to measure mercury in paint (Anonymous
1972).

TABLE A.3 Methods of Analysis

(1) Analytical Method	(2) Important Application	(3) State of Measured Mercury	(4) Sample Preparation	(5) Methodology or Technique
Gas chromatography using electron detector	Biological material; Waste sludge: Bottom mud	Organomercury Components	Sample is homogenized; treated with acid, organic solvents, etc. Components are extracted with benzene.	Sample components are separated by chromatographic technique and analyzed qualitatively and quantitatively.
Atomic absorption spectroscopy cold vapor method	Effluents, wastewaters, surface and saline waters; soils and sediments after dissolution in aqua regia. Also for air and biological materials.	Elemental mercury vapor	Acidify sample with HNO_3 at time of collection. Filter if necessary, then chemically convert all mercury to mercuric ion.	Mercury is reduced to elemental state then aerated into a cell placed in the path of light of 253.7 nm wavelength. The measured absorbance is a function of mercury concentration.
Neutron activation analysis	All media and materials, biologic and inorganic.	As is	Water samples must be concentrated or extracted onto an ion exchange bed. Other materials are dried or freeze dried.	Sample is irradiated for 1-24 hours with a stream of neutrons. 197 Hg and 203 Hg are counted using a multichannel analyzer.
Colorimetric spectrophotometry	Aqueous samples, biological materials; soils and sediments.	Mercury dithizone complex	Sample is shaken with excess dithizone in acidic solution then extracted with organic solvents (organomercurials require oxidative digestion).	Absorbance of the complex at 490 nm or reduction in absorbance at 610 nm is measured.

TABLE A.3 (Continued)

(6) Reported Sensitivity	(7) Precision (relative standard deviation/sample size)	(8) Accuracy (recovery from spiked sample)	(9) Some Interfering Substances	(10) Selectivity	(11) Comments
0.01 μg/g (fish)	±10% (0.3 μg/g)	95.5% (0.20 μg/g)	Sulfide in sample requires refired pretreatment to free organomercury.	The concentration and species of organomercury compound is determined.	Loss of volatile mercury compounds during pretreatment must be avoided. Injection system should be checked to prevent alteration of the organomercury compound in the injection chamber.
0.001 μg/g (sediment)	±10% (0.01 μg/g)	96.3% (0.01 μg/g)			
0.20 μg/g	±46% (0.55 μg/g)	89%	Sulfide, free chlorine, iodine, CO_2, SO_2, and volatile organics interfere. Copper may interfere.	Total mercury concentration is determined.	Mercury vapor must be dried before passing through the absorption cell. Water vapor absorbs at 253.7 nm.
0.0005 μg/g (involves post-irradiation separation)	±5% (0.1 μg/g)	98%	Sodium, bromine, phosphorus may interfere. Separation techniques avoid the interference.	Total mercury is determined.	Care must be taken to prevent mercury loss by volatilization during irradiation.
0.05-0.5 μg/g	4-5%		Copper, silver, gold, palladium and platinum in trace amounts.	Total mercury is determined.	Good, but lengthy method for routine analysis of samples containing 0.5 - 50 μg/g Hg and only traces of Pb, Zn, Ni or Co.

SOURCES: U.S. EPA (1973, 1974), Anon. (1973a), Friberg and Vostal (1972), Wallace et al. (1971).

Volumetric Analysis

Aqueous volumetric procedures depend upon the conversion of mercury to Hg^{2+} ion, complexation of the ion, then titration to a readable end point. In one widely used method, the Volhard titration, potassium thiocyanate is added in excess to form the sparingly soluble compound, mercuric thiocyanate. The excess thiocyanate is titrated with ferric ion to produce a brilliant red color. The advantage of the Volhard titration is that very few metals interfere; the disadvantage is that the reaction is not entirely stoichiometric. Thus, corrections are needed that may introduce uncertainties beyond the accepted tolerances for some trace analyses (Coetzee 1961). There are also other titrimetric reactions that involve oxidation reduction reactions.

Micrometric Determination

With this method, mercury is electrodeposited from solution onto copper wire; the mercury is then distilled from the copper wire in a capillary tube. The volatilized mercury condenses in a cooled section of the tube, the condensate is united into a globule, and the diameter of the globule is measured with a microscope. Under favorable conditions, amounts exceeding 0.5 μg can be determined with an accuracy of ±2 percent, and as little as 0.01 μg with an accuracy of ±10 percent (Coetzee 1961).

Colorimetric Spectrophotometry

Until the development of more sensitive instrumental methods, dithizone extraction followed by photometric analysis was the method of choice for trace environmental mercury samples. Dithizone (diphenylthiocarbazone) complexes quantitatively with mercuric ion in acidic aqueous solution. The complex is extracted with chloroform as orange mercury dithizonate, which exhibits an absorption maximum at 490 nm. Various techniques are required to eliminate interfering metals (Sandell 1959). The sensitivity and procedures of the method are outlined in Table A.3.

Cold Vapor Atomic Absorption Spectrometry

The use of a cold vapor atomic absorption method for trace mercury analysis is so direct and accessible under

ordinary laboratory conditions that it has been detailed
as an experiment for instruction in an undergraduate
course in analytical chemistry (Lieu et al. 1974). The
equipment is shown in Figure A.1. A hollow cathode
mercury discharge lamp supplies radiation at 253.7 nm
for absorption by atomic mercury vapor aerated from
solution into the absorption cell. Mercury ions in
solution are reduced to elemental mercury by stannous
chloride, then swept with helium into the cell where
radiation is absorbed by the mercury atoms. The
concentration of atoms is determined from the absorbance
by comparison with a standard curve. Atomic absorption
is highly specific for the particular metal under
investigation. Most interfering agents are readily
removed, and it is suitable for measurement of most
environmental materials after pretreatment. Elemental
mercury in air, for instance, is monitored directly by
the passage of air through an atomic absorption cell
(Jepsen 1973). Unfortunately, atomic absorption
determines only the total mercury content and gives no
information about the species or form of the mercury.

The cold vapor method has been readily adopted and
will likely become the favored method eventually. It is
fast, specific for mercury, suitable for all sorts of
pretreatment variations, applicable to extremely low
concentrations, and above all, the equipment is readily
available and reasonable in cost. The Bureau
International Technique Du Chlor has selected cold vapor
atomic absorption spectrometry as their standard
(Anonymous 1974). The sensitivity and accuracy are well
within the 0.1 $\mu g/g$ ±10 percent cited as desirable for
regulatory purposes (Alvarez 1974).

Gas Chromatography with an Electron Capture Detector

Gas chromatography actually refers to a separation
technique. Coupling this with a detector for sensing
the separated components constitutes an instrumental
method, the only one in wide use that determines organic
mercury compounds directly. These compounds are
converted to organomercuric halides to make the mercury
volatile and chromatographable. After extraction into
an organic solvent, the organic mercury compounds are
injected into an inlet where they are vaporized and the
gases passed through a chromatographic column. The
absorbent in the column is often a high-boiling liquid
adsorbed on a solid substrate; hence, the term gas-
liquid chromatography. The volatile organics,
partitioned between the stationary phase and the mobile
carrier gas, emerge from the column at different times
and are determined by one of several detection devices.

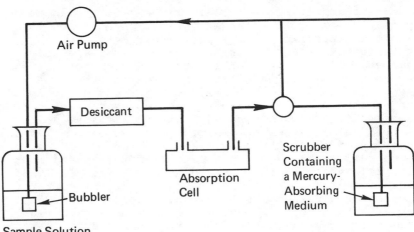

Air Pump

Desiccant

Bubbler

Sample Solution
in BOD Bottle

Absorption
Cell

Scrubber
Containing
a Mercury-
Absorbing
Medium

SOURCE: U.S. EPA (1974).

FIGURE A.1   Apparatus for flameless mercury determination.

The electron capture detector, used almost exclusively in organomercury determination by gas chromatography, employs a tritium or [63]Ni radioactive source. Its beta radiation produces electrons in the carrier gas that are captured by sample molecules, thus reducing the current of electrons to an anode under fixed voltage. The decrease in current is a measure of the amount and electron affinity of the component (Andelman 1971).

Dressman (1972) found that phenylmercuric salts, except phenylmercuric chloride, are converted in the gas chromatographic injection block to diphenylmercury as the major product, which is overlooked by electron capture detection. Baughman et al. (1973) found a similar conversion of methylmercury salts to dimethylmercury, and they suggest the use of specially treated columns for reliable analyses.

Neutron Activation Analysis

The use of neutron activation analysis of mercury in bioenvironmental materials is now well established. Usually, the sample is sealed in quartz vials and irradiated with neutrons to convert [196]Hg to [197]Hg, a radioactive isotope with a 65 h half-life. The [197]Hg is identified with a multichannel analyzer and Ge(Li) detector (Westermark 1972). The most important and useful aspect of activation analysis is that no treatment of the sample is needed before neutron irradiation, although postirradiation separation can be performed to increase sensitivity by removal of interfering substances.

The major limitation on activation analysis is its inability to distinguish the chemical state of mercury in the sample. In addition, the excessive equipment cost relative to other methods and the need for a nearby neutron source (reactor) has, to some extent, reserved this method primarily for referee analyses.

Newer Methods of Analysis

Atomic Fluorescence Spectrometry

Atomic fluorescence depends on excitation of mercury atoms with a mercury lamp followed by detection of the emitted radiation when the atoms return to the ground state. This method is reported to have a sensitivity advantage of at least a factor of 10 over atomic absorption (Subber et al. 1974). Atomic fluorescence spectrometry is useful to detect mercury levels between

0.1 ng and 1 μg. Above 1 μg the calibration curve becomes nonlinear (Subber et al. 1974).

The techniques for pretreatment and for reduction and release of mercury from solution are exactly the same as for atomic absorption. Here, however, the 253.7 nm radiation from a mercury line source is focused on the vapor, and the resulting 253.7 nm fluorescence emission is monitored. The added sensitivity is derived, in part, from the fact that the absorption is a very small quantity determined as the difference between two large quantities, whereas emission is this same quantity determined directly (Minear and Murray 1974). In addition, the emission is a function of the intensity of the impinging radiation, which can be varied to some extent to increase the signal. Muscat et al. (1972) used atomic fluorescence to determine mercury in a number of environmental reference materials and sediments. For a water sample from the Analytical Quality Control Laboratory, Cincinnati, containing 4.2 ng/ml, a value of 4.2±0.4 ng/ml was obtained. Wheat flour samples (International Atomic Energy Agency, Code 66/10) yielded 5.1±0.5 μg/g compared with a pooled analysis of 4.59±1.32 μg/g by neutron activation, or 4.92±0.45 μg/g by atomic absorption. Atomic fluorescence appears to be a promising technique in terms of sensitivity, availability, and ease of operation.

Spark Source Mass Spectrometry

In an analytical program undertaken by the NBS to provide a reliable value for the mercury content of ground orchard leaves, a standard environmental reference material (SRM 1571), three analytical methods were used: atomic absorption spectrometry, neutron activation, and stable-isotope dilution with spark source mass spectrometry (Alvarez 1974). The first two are well tested, reliable methods. Spark source mass spectrometry has not been previously applied to environmental mercury problems, although it has been used at NBS for standardization needs. The apparatus consists of a spark source mass spectrometer with Mattauch-Herzog geometry (double-focusing). The environmental standards were processed routinely to oxidize the organic matter; then the mercury-containing solution was spiked with 0.370 μg ^{201}Hg and 50 μg ^{198}Hg as a carrier. The mercury was electrodeposited onto high-purity gold wires for sparking in the mass spectrograph. The concentration was calculated from the altered isotope ratio, ^{201}Hg/^{202}Hg. The results are given in Table A.4. The average concentration,

0.141±0.009 μg/g, is in good agreement with averages determined not only by other methods but by other laboratories.

Gas Chromatography with a Microwave Emission Spectrometric Detector

A gas/chromatography microwave-excited-spectrometric detector (GC-MES) has been built and applied to detect dimethylmercury, diethylmercury and methylmercuric chloride at the 6-, 7-, and 4-pg levels, respectively (Talmi 1974). The relative sensitivity is at the 5 ng/g level. The technique is reported to be substantially faster and simpler than most that are presently available. In this detector, the intensity of the 253.7-nm line emitted from the microwave generated plasma used as a GC detector is monitored to characterize the molecule and to determine its concentration. A comparison with a conventional electron capture detector (ECD) is given in Table A.5. The GC-MES appears to be equal to the GC-ECD in sensitivity, yet superior in all other respects.

Gas Chromatography with a Mass-Spectrometric Detector

Although gas chromatography/mass spectrometry has been applied for at least 5 yr to study various organic pollutants (Webb et al. 1973), it has been used only sparingly to detect organomercural compounds (Johansson et al. 1970). Baughman et al. (1973) used a gas-liquid-chromatography/mass-spectrometry (GLC-MS) apparatus to demonstrate that ionic methylmercury compounds undergo decomposition during GLC. Johansson et al. (1970) analyzed eight samples of fish flesh containing from 0.14 to 3.2 μg/g mercury as methylmercury. Excellent agreement was obtained in a comparison with the results from using gas chromatography/mass spectrometry, gas chromatography/electron capture detection, and neutron activation analysis. The three methods deviated less than ±10 percent from the average value. The mass spectrometer provides a positive identification of organomercury compounds, including organomercuric iodide.

Stripping Voltammetry

Anodic stripping voltammetry is a convenient, sensitive method for metals analyses in aqueous media, particularly seawater (Smith and Windom 1972). The

TABLE A.4 Comparison of Mercury Results by Isotope Dilution with Other Analytical Methods for Orchard Leaves, SRM 1571

Average Concentration and 95% Confidence Limits (μg/g)	Number of Determinations	Analytical Method	Lab
0.141 ± 0.009	6	Isotopic dilution	NBS
0.160 ± 0.012	15	Atomic absorption	NBS
0.155 ± 0.006	11	Neutron activation	NBS
0.145 ± 0.014	5	Neutron activation	A
0.148 ± 0.010	4	Neutron activation	B

SOURCE: Alvarez (1974).

TABLE A.5 Comparison Between Electron Capture and Plasma Emission Spectroscopy Systems

Item	Electron Capture	Plasma Emission Spectroscopy
Selectivity	Response to electron-absorbing compounds, especially halogens, nitrates, and conjugated carbonyls	Spectroscopical selectivity. Sensitive to practically all elements. Can be used both as selective and general-purpose detector
Sensitivity	Picogram level	Picogram level
Linear range[a]	50 to 100	10^3 to 10^4
Stability	Fair	Good to excellent
Temperature limit	250°C (^{3}H), 350°C (^{63}Ni)	Practically unlimited
Carrier gas	N_2 or Ar– 10% CH_4	Ar and He
Repairs	By producer only	Very simple procedure; usually involves a replacement of the quartz capillary
Life expectancy	Limited by "poisoning" compounds	Practically unlimited; unaffected by any compounds
Remarks	Easily contaminated, easily cleaned, sensitive to water, carrier gas must be dried	Not contaminated, very easy to clean, insensitive to water, sensitive to nitrogen tracers

[a]Linear range is defined as the ratio of the highest to the lowest concentration values that lie on a linear calibration curve.

SOURCE: Talmi (1974).

customary electrode, a mercury drop or mercury film, however, is obviously unsuitable for mercury determination; and since mercury amalgamates with platinum or gold electrodes, another electrode material is required. Allen and Johnson (1973) used a rotating ring-disk electrode that had a glassy carbon disk plated with a thin film of gold. Mercurous ion in 1.0 M sulfuric acid solution was determined in the 0.10- to 4.00-μg/l range, with a relative standard deviation of 7.5 percent. The limit of detection for the technique is approximately 0.01 μg/l. This method is academic and will probably never be used in actual analyses for mercury.

COMPARISON OF ANALYTICAL METHODS

Gas chromatography is generally accepted as the most sensitive, accurate method for organomercury determinations. Controversy continues, however, over the relative merits of atomic absorption spectroscopy and neutron activation analysis for total mercury determinations. Wood (1972) claims the latter technique is superior, asserting that consistently lower results are obtained with atomic absorption analyses. Hume (1973), however, cites a comparison of analytical results for cobalt in two samples of seawater from eight laboratories, four that use atomic absorption with chelation-extraction and four that use neutron activation. The absorption averages were 5 to 6 times larger than those for activation.

Interlaboratory comparisons of different instrumental methods are necessary to elucidate the advantages and pitfalls of various techniques and to display the uncertainty inherent in some trace analyses.

Standardization

Analyses for trace levels of mercury are performed in laboratories throughout the world. Comparisons of these results are difficult, however, because of a lack of uniformity. Kaiser (1973) has addressed this problem in terms of trace analyses in general, presenting an excellent treatment of expressions of accuracy as mentioned above in the section on analytical procedure.

All values measured are subject to errors, e.g., systematic and random. The random error is usually evaluated by the individual investigator or laboratory from a statistical analysis of the data. Systematic errors, on the other hand, are not readily accessible to evaluation individually, although blank readings are intended to eliminate them. Moreover, this type of error acquires additional importance in trace element determinations because it may contribute disproportionately to the "analytical signal." Wilson (1974) has reviewed the role of systematic and random errors in standardization procedures. An acceptable test for systematic errors includes the use of biological and environmental standard reference materials. They enable the analyst to verify the accuracy (absence of systematic error) in his analytical procedure. The NBS supplies certified biological and environmental samples tested by many techniques, the results of which are shown in Table A.4 for Orchard Leaves, SRM 1571 (Alvarez 1974).

Interlaboratory Comparisons

Interlaboratory comparisons have been conducted since the mid 1960s. These have involved not only different laboratories but also different instrumental methods. An examination of the results of a few of these studies indicates a trend toward increased accuracy, although it is not always immediately apparent, since comparisons have shifted from the $\mu g/g$ to the ng/g range.

A comparative analysis of a standard plant material submitted to laboratories around the world before 1967 resulted in an inconsistent average mercury content of 0.150 ± 0.008 $\mu g/g$ by activation analysis and 0.0122 ± 0.0024 $\mu g/g$ by colorimetric analysis (Bowen 1967). Rottschafer et al. (1971) likewise reported the following inconsistent results when replicate fish tissue samples were analyzed for mercury in 1970 or earlier by different laboratories and various techniques:

Analytical Method	Results ($\mu g/g$)
Atomic absorption, laboratory 1	0.06
Atomic absorption, laboratory 2	1.00
Atomic absorption, laboratory 3	0.85
X-Ray fluorescence	0.40
Destructive neutron activation	0.87

A study reported in 1974 is shown in Table A.4. Here, not only was excellent agreement obtained between different laboratories with the same technique, but also by one laboratory with three different methods. Note that the 100-$\mu g/g$-concentration range was chosen as desirable for regulatory purposes (Alvarez 1974). Another recent study (Anonymous 1974) involved one method, atomic absorption spectroscopy, 37 laboratories, and samples containing mercury that varied over 3 orders of magnitude of concentration from 20 $\mu g/g$ to 20 ng/g. The results are shown in Table A.6. The relative standard deviation from the mean is plotted against the concentration in Figure A.2 to show the strong functional dependence of the reproducibility on concentration. High relative standard deviations occurred at low concentrations, but such results are not unexpected insofar as "analytical noise" is concerned (Kaiser 1973).

The studies cited here are not intended to be a comprehensive treatment of interlaboratory comparisons with regard to analytical capability. Rather, they are meant to demonstrate a definite trend toward improved precision and accuracy in procedures for analyzing mercury in the environment. Currently, mercury and organomercurials can be determined about as well as any trace element or trace organometallic of environmental significance (Shults 1975).

NOTE

1 This appendix is adapted from pages 26-59, Review of the Environmental Effects of Mercury, prepared by the Biomedical Sciences Section, Information Center Complex, Information Division, Oak Ridge National Laboratory for the U.S. Environmental Protection Agency, EPA-IAG-D4-0403. January 27, 1975.

TABLE A.6 Results of Statistical Evaluation of Interlaboratory Analysis for Mercury

| Item | Standard Samples | | Waste Water | Caustic Soda | |
	Ctg. 5.0 mg Hg/kg	Ctg. 20.0 mg Hg/kg		Sample 1	Sample 2
Arithmetic mean	5.03 mg/kg	20.3 mg/kg	3.18 mg/kg	21.4 µg/kg	201 µg/kg
Repeatability[a]: Standard deviation	0.18 mg/kg	0.8 mg/kg	0.10 mg/kg	1.7 µg/kg	11 µg/kg
Relative standard deviation	3.5%	4.0%	3.1%	8.0%	5.4%
Reproducibility[b]: Standard deviation	0.30 mg/kg	1.1 mg/kg	0.34 mg/kg	4.8 µg/kg	38 µg/kg
Relative standard deviation	5.9%	5.4%	11%	22%	19%

[a]Repeatability means single laboratory, single operator, and single apparatus precision;
[b]Reproducibility means multi-laboratory, multi-operator, and multi-apparatus precision.

SOURCE: Anon. (1974).

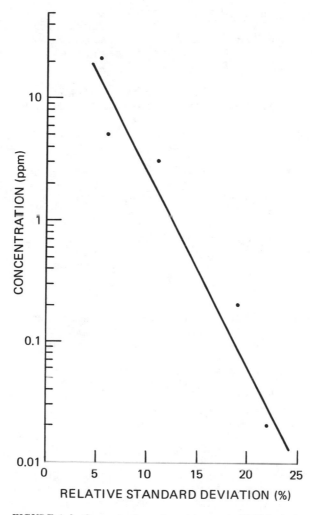

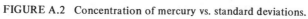

FIGURE A.2 Concentration of mercury vs. standard deviations.

REFERENCES

Allen, R.E. and D.C. Johnson (1973) Determination of Hg(II) in acidic media by stripping voltametry with collection. Talanta 20:799-809.

Alvarez, R. (1974) Sub-microgram per gram concentrations of mercury in orchard leaves determined by isotope dilution and spark-source mass spectrometry. Anal. Chim. Acta 73:33-38.

Andelman, J.B. (1971) Concentration and separation techniques. Pages 33-62, Instrumental Analysis for Water Pollution Control, edited by K.H. Mancy. Ann Arbor, Mich.: Ann Arbor Science Publishers.

Anonymous (1972) Annual Book of ASTM Standards, Part 21. Philadelphia, Pa.: American Society for Testing and Materials.

Anonymous (1973a) Instrumentation for Environmental Monitoring--Biomedical, Lawrence Berkeley Laboratory, University of California. Berkeley, Calif.: University of California.

Anonymous (1973b) Catalog of Standard Reference Materials, NBS Special Publication 260. Washington, D.C.: U.S. Government Printing Office.

Anonymous (1974) Standardization of methods for the determination of traces of mercury. Part I. Anal. Chim. Acta 72:37-48.

Baughman, G.L., M.H. Carter, N.L. Wolfe, and R.G. Zepp (1973) Gas-liquid chromatography-mass spectrometry of organomercury compounds. J. Chromatogr. 76:471-476.

Becknell, D.E., R.H. Marsh, and W. Allie, Jr. (1971) Use of anion exchange resin-loaded paper in the determination of trace mercury in water by neutron activation analysis. Anal. Chem. 43:1230-1233.

Bogen, J. (1973) Trace elements in atmospheric aerosol in the Heidelberg area measured by instrumental neutron activation analysis. Atoms. Environ. 7:1117-1125.

Bowen, H.J.M. (1967) Comparative elemental analyses of a standard plant material. Analyst 92:124-131.

Campbell, E.E., G.O. Wood, P.E. Trujillo, and P. Stein (1972) Evaluation of Methods for Determining Mercury. LASL Project R-060, LA-5188-PR. Los Alamos Scientific Laboratory of the University of California, Los Alamos, N.M.

Coetzee, J.F. (1961) Mercury. Pages 231-326, Treatise on Analytical Chemistry, Part II, Vol. 3, edited by I.M. Kolthoff, P.J. Elving, and E.B. Sandell. New York: Interscience Publishers.

Doherty, P.E. and R.S. Dorsett (1971) Determination of trace concentrations of mercury in environmental water samples. Anal. Chem. 43:1887-1888.

Dressman, R.C. (1972) The conversion of phenylmercuric salts to diphenylmercury and phenylmercuric chloride upon gas chromatographic injection. J. Chromatogr. Sci. 10:468-472.

Feldman, C. (1974) Preservation of dilute mercury solutions. Anal. Chem. 46:99-102.

Fishman, M.J. (1970) Determination of mercury in water. Anal. Chem. 42:1462-1463.

Friberg, J. and J. Vostal (eds.) (1972) Mercury in the Environment--A Toxicological and Epidemiological Appraisal. Cleveland, Ohio.: CRC Press.

Hume, D.N. (1973) Pitfalls in the determination of environmental trace metals. Pages 3-16, Chemical Analysis of the Environment, edited by S. Ahuja, E.M. Cohen, T.J. Knelp, J.L. Lambert, and G. Zeveig. New York: Plenum Press.

Jepsen, A.F. (1973) Measurements of mercury vapor in the atmosphere. Pages 81-95, Trace Elements in the Environment, edited by E.L. Kothny. Advances in Chemistry Series 123. Washington, D.C.: American Chemical Society.

Johansson, B., R. Ryhage, and G. Westoo (1970) Identification and determination of methylmercury compounds in fish using combination gas chromatograph - mass spectrometer. Acta Chem. Scand. 14:2349-2354.

Kaiser, H. (1973) Guiding concepts relating to trace analysis. Pages 35-61, Analytical Chemistry - 4. International Union of Pure and Applied Chemistry. New York: Crane, Russak and Co.

Kopp, J.F., M.C. Longbottom, and L.B. Lobring (1972) Cold vapor method for determining mercury. J. Amer. Water Works Assoc. 64:20-25.

Lieu, V.T., A. Cannon, and W.E. Huddleston (1974) A non-flame atomic absorption attachment for trace mercury determination. J. Chem. Educ. 51:752-753.

Linch, A.L. R.F. Stalzer, and D.T. Leffers (1968) Methyl and ethyl mercury compounds--recovery from air and analysis. Amer. Indust. Hyg. Assoc. J. 29:79-86.

Lindstedt, G. and S. Skerfving (1972) Methods of analysis. Pages 3-13, Mercury in the Environment, edited by L. Friberg and J. Vostal. Cleveland, Ohio: CRC Press.

Longbottom, J.E., R.C. Dressman, and J.J. Lichtenberg (1973) Gas chromatorgaphic determination of methyl mercury in fish, sediment and water. J. Assoc. Offic. Anal. Chem. 56:1297-1303.

Minear, R.A. and B.B. Murray (1974) Methods of trace metals analysis in aquatic systems. Pages 1-42, Trace Metals and Metal Organic Interaction in Natural Waters, edited by P.C. Singer. Ann Arbor, Mich.: Ann Arbor Science Publishers.

Moffitt, A.E., Jr. and R.E. Kupel (1970) A rapid method employing impregnated charcoal and atomic absorption spectrophotometry for the determination of mercury in atmospheric, biological and aquatic samples. Atomic Absorption Newsletter 9:113-118.

Muscat, V.I., T.J. Vickers, and A. Andren (1972) Simple and versatile atomic fluorescence system for determination of nanogram quantities of mercury. Anal. Chem. 44:218-221.

Mushak, P. (1973) Gas-liquid chromatography in the analysis of mercury (II) compounds. Environmental Health Perspectives 4:55-60.

Riley, J.P. and D. Taylor (1968) Chelating resins for the concentration of trace elements from sea water and their analytical use in conjunction with atomic absorption spectrophotometry. Anal. Chim. Acta 40:479-485.

Rosain, R.M. and C.M. Wai (1973) The rate of loss of mercury from aqueous solution when stored in various containers. Anal. Chim. Acta 65:279-284.

Rottschafer, J.M., J.D. Jones, and H.B. Mark, Jr. (1971) A simple, rapid method for determining trace mercury in fish via neutron activation analysis. Environ. Sci. Technol. 5:336-338.

Sandell, E.B. (1959) Colorimetric Determination of Metals, 3rd Ed. New York: Interscience Publishers, Inc.

Shults, W.D. (1975) Personal communication, Analytical Chemistry Division, Oak Ridge National Laboratory, Oak Ridge, Tenn. January.

Smith, R.G., Jr. and H.I. Windom (1972) Analytical Handbook for the Determination of Arsenic, Cadmium, Cobalt, Copper, Iron, Lead, Manganese, Mercury, Nickel, Silver and Zinc in the Marine and Estaurine Environments. Scidaway Institute of Oceanography, Savannah, Ga., University of Georgia. (Unpublished manuscript.)

Subber, S.W., S.D. Fihn, and C.D. West (1974) Simplified apparatus for the flameless atomic fluorescence determination of Hg. Amer. Lab. 6:38-40.

Talmi, Y. (1974) Separation, detection and other identification of organically bound toxic metals and other hazardous materials. Ecology and Analysis of Trace Contaminants, edited by W. Fulkerson, W.D. Shults, and R.I. Van Hook. ORNL-NSF-EACT-6. Oak Ridge, Tenn.: Oak Ridge National Laboratory.

Thatcher, L.L. and J.O. Johnson (1971) A comprehensive program in neutron activation analysis in water quality. Pages 323-328, Nuclear Techniques in Environmental Pollution. Vienna: International Atomic Energy Agency.

Trujillo, C., P. Stein, and E. Campbell (1974)
Preservation of dilute mercury solutions. Anal. Chem.
46:99-102.

U.S. Environmental Protection Agency (1972) Handbook of
Analytical Quality Control in Water and Wastewater
Laboratories. Analytical Quality Control Laboratory,
U.S. Environmental Protection Agency. Cincinnati,
Ohio: U.S. Environmental Protection Agency.

U.S. Environmental Protection Agency (1973) Proposed
Water Quality Information, Vol. II. Washington, D.C.:
U.S. Environmental Protection Agency.

U.S. Environmental Protection Agency (1974) Manual of
Methods for Chemical Analysis of Water and Wastes.
National Environmental Research Center. EPA-625/6-74-
003. Cincinnati, Ohio: U.S. Environmental Protection
Agency.

Uthe, J.F. (1971) Determination of total and organic
mercury levels in fish tissue. Pages 207-212,
Identification and Measurement of Environmental
Pollutants. National Research Council of Canada,
Ottawa, Canada: Campbell Printing.

Wallace, R.A., W. Fulkerson, W.D. Shultz, and W.S. Lyon
(1971) Mercury in the Environment. ORNL-NSF-EP-1. Oak
Ridge, Tenn.: Oak Ridge National Laboratory.

Webb, R.G., A.W. Garrison, L.H. Keith, and J.M. McGuire
(1973) Current Practices in GC-MS Analysis of Organics
in Water. National Environmental Research Center.
Corvallis, Oreg.: U.S. Environmental Protection
Agency. 91 pp.

Weiss, H.V. and K. Chew (1973) Neutron irradiation of
mercury in polyethylene containers. Anal. Chim. Acta
67:444-447.

Westermark, T. (1972) Activation analysis of mercury in
environmental studies. Pages 57-88, Advances in
Activation Analysis, Vol. 2. New York: Academic Press.

Westoo, G. (1966) Determination of methylmercury
compounds in foodstuffs. I. Methylmercury compounds in
fish, identification and determination. Acta Chem.
Scand. 20:2131-2137.

Westoo, G. (1967) Determination of methylmercury
compounds in foodstuffs. II. Determination of
methylmercury in fish, egg, meat and liver. Acta Chem.
Scand. 21:1790-1800.

Westoo, G. (1968) Determination of methylmercury salts
in various kinds of biological media. Acta Chem.
Scand. 22:2277-2280.

Wilson, A.L. (1974) Performance characteristics of
analytical methods. IV. Talanta 21:1109-1121.

Wood, J.M. (1972) A progress report on mercury.
Environment 14:33-39.

A BRIEF REVIEW OF FAO/WHO DELIBERATIONS
ON SETTING ACCEPTABLE TOLERANCES
FOR MERCURY RESIDUES IN FOODS, 1963-1976[1]

In a series of joint meetings between 1963 and
1976 the FAO Working Party on Pesticide Residues
and the WHO Expert Committee on Pesticide Residues
have consistently taken the position that no firm
basis has yet been established for determining
what would be a safe level of mercury in food to
protect the health of human beings. Therefore, no
acceptable daily intake (ADI) or tolerance levels
have yet been established despite suggested
recommendations for provisional standards, the
most recent in 1972.
Nevertheless, a great deal of confusion continues with
respect to an ADI and/or food tolerance for total
mercury allegedly set in 1963 when the First Session of
the FAO/WHO Codex Alimentarius Commission convened to
implement the Joint FAO/WHO Food Standards Program.
They were provided with an Extract from the Draft Latin-
American Food Code, which is appended to their report as
Appendix E.2 (FAO/WHO 1963). Article 10 of the Draft
reads, in part, as follows:

> The presence of the metals and metalloids
> (incidental or residual additives) listed
> hereinafter shall be tolerated in foods (with the
> exception of drinking water, fish and shellfish),
> provided that they are kept within the following
> limits: Mercury...Maximum: 0.05 parts per million.

While this was given a "first reading'' by the First
Session of the Codex Alimentarius Commission, it was not
adopted and should not be interpreted to mean that the
commission accepted a level of 0.05 μg Hg/g in
foodstuffs (D.G. Chapman, personal communication, Food
Safety Unit, World Health Organization, Geneva, 1977).
Nevertheless, erroneous reports of its adoption continue
(Goldwater 1974). Later in 1963, on the basis of animal
studies, the FAO/WHO Joint Committee on Pesticide
Residues estimated but did not set or recommend a

maximum acceptable daily intake of 0.05 µg per kg of body weight phenylmercuric acetate for humans (FAO/WHO 1964).

The 1966 FAO Working Party and WHO Expert Committee on Pesticide Residues again considered the toxicology of a number of organomercurial fungicides used in agriculture and once more concluded that sufficient data were not available to arrive at an acceptable daily intake (ADI) or even a temporary tolerance. By way of guidance, however, they suggested a practical residue limit of from 0.02 to 0.05 µg Hg/g of food according to local conditions (FAO/WHO 1967a). This practical residue limit was also suggested in 1967 but has not been included in any of the subsequent Joint FAO/WHO reports on pesticide residues (FAO/WHO 1968a). The 1967 report also noted that small natural concentrations of mercury are widespread, but the levels vary from one area to another. Moreover, where crops or foodstuffs are treated with organomercurials in accordance with good agricultural practices, the following maximum levels of mercury residues were found: rice, 0.3 µg/g (provisional); apples and tomatoes, 0.1 µg/g; eggs and meat (except liver and kidney), 0.1 µg/g; potatoes, 0.05 µg/g; and wheat and barley, 0.03 µg/g (FAO/WHO 1968a).

The 1966 FAO/WHO Expert Committee on Pesticide Residues and Food Additives also attempted to compute the possible daily intakes of pesticide residues on foods based on high estimates of food consumption in the United States (FAO/WHO 1967b, 1967c). To do this, they assumed that all foodstuffs within the same category were treated with the pesticide. Thus, the high consumption value for fish and shellfish was estimated at 60 g per person per day. The average daily dietary intakes of mercury were estimated to range from 0.3 to 1.0 µg per kg body weight (21 to 70 µg per 70-kg person per day, 0.15 to 0.49 mg per 70-kg person per week), with little accumulation in the tissues. These committees recommended that every effort be made to control and reduce mercury contamination of the environment and consequently the levels in food. They also affirmed the urgent need for further studies of the distribution of mercury in foods and beverages and in human tissues of various ages in different environments.

In 1967 the Joint Meeting of the FAO Working Party and the WHO Expert Committee on Pesticide Residues again set no ADI, tolerance or practical residue limits because of a lack of data. However, they strongly discouraged agricultural uses that would increase the level of mercury in food. Nonetheless, the committee contended that organomercurial seed dressings applied up to the time of petal fall would not result in residues on apples in excess of normal background levels (FAO/WHO

1968a, 1968b). This position was continued until 1969 (FAO/WHO 1969, 1970). After 1967 the succeeding FAO/WHO Committees on Pesticide Residues did not directly recommend a practical residue limit but until 1972 did refer to the suggested practical residue limit of 0.02 to 0.05 μg Hg per g originally published in 1967 (FAO/WHO 1968a).

The 1969 FAO/WHO report on pesticide residues in food noted the wide use of organomercurials to protect seed and seed grain and the high toxicity of these compounds to human beings, as well as the fact that unintentional residues were appearing in food crops, animals, and animal products. Therefore, the report stressed the need for safer substitutes and urged governments and other concerned organizations to give this research a high priority (FAO/WHO 1970).

In April 1972, the Joint FAO/WHO Expert Committee on Food Additives (FAO/WHO 1972b, 1972c) finally proposed a provisional tolerable intake (PTI) of 0.3 mg mercury per person per week, of which no more than 0.2 mg should be in the methylated form. For a 70-kg adult, this is equivalent to a daily intake of about 0.6 μg per kg of body weight, with no more than 0.4 μg per kg per day as methylmercury. The 1972 Joint FAO/WHO Expert Committee on Food Additives also stated that the use of alkyl and aryl mercurial fungicides as seed dressings should be discouraged. They noted that aryl and inorganic mercury compounds can be biologically converted into alkylmercury compounds by organisms in nature and recommended research to replace the organomercurial fungicides with compounds that are less likely to poison human beings accidentally. This recommendation was based on the committee's acknowledgement that a number of fatal human poisonings had occurred because seeds treated with alkylmercury compounds had accidentally or improperly been diverted from sowing to human consumption. Nevertheless, the 1972 committee again confirmed the opinion of the 1967 committee (FAO/WHO 1968a) that the uptake of mercury into crops from seed dressing was insignificant as a potential source of food contamination. They recommended that the uses of mercury compounds in agriculture be reviewed by the FAO Working Party of Experts on Pesticide Residues and the WHO Expert Group on Pesticide Residues (FAO/WHO 1973).

In 1973 the FAO/WHO Codex Alimentarius Commission (1974a) noted that no limits were set for mercury in food, nor was a Codex referee method of analysis established. However, for human beings a provisional tolerable weekly intake (PTWI) was set at 5 μg per kg body weight for total mercury intake and 3.3 μg per kg body weight for methylmercury expressed as mercury.

While all mention of mercury fungicides has been deleted from the FAO/WHO Joint Report of Pesticide Residues in Food since 1973 (FAO/WHO 1974c, 1975, 1976), special publications have continued to deal with the subject. The FAO/WHO (1974b) Joint Report on The Use of Mercury and Alternative Compounds as Seed Dressings contended that until effective and broad spectrum alternative fungicides are available, the use of alkoxyalkylmercury and arylmercury compounds will have to continue under safeguards decided upon by individual countries. In addition, in light of the serious poisoning outbreaks in various countries, the committee recommended that alkylmercurials be limited to the treatment of nuclear stocks in the first few generations of seed multiplication and that the handling of the treated seed be strickly controlled. Because of the known high toxicity of alkylmercury compounds, it was also recommended that their use never be permitted for the treatment of seed to be exported for the production of food. The following year the World Health Organization sponsored a conference on human intoxication resulting from alkylmercury-treated seeds and dealt mainly with the Iraqi methylmercury poisoning outbreak of 1971-1972 (WHO 1976a). As new data were added about this condition, the resolve was reinforced to prevent similar outbreaks at all cost in the future. Finally, in 1976 the World Health Organization published Environmental Health Criteria 1, Mercury (WHO 1976b) assessing not only the existing information on the relationship between mercury exposure and man's health but also providing guidelines for setting exposure limits consistent with health protection.

NOTE

1 For definitions of terms accepted by the FAO/WHO Joint Meetings on Pesticide Residues in Food see:

- FAO/WHO (1976) for definitions of pesticide, pesticide residue, good agricultural practice in the use of pesticides, acceptable daily intake, temporary acceptable daily intake, conditional acceptable daily intake, potential daily intake, maximum residue limit, temporary maximum residue limit, extraneous residue limit, further work required;
- FAO/WHO (1970) for definitions of practical residue limit, tolerance, and temporary tolerance; and
- FAO/WHO (1972b) for definition of provisional tolerable weekly intake.

REFERENCES

FAO/WHO (1963) Report of the First Session. Joint
FAO/WHO Codex Alimentarius Commission, Rome, June 25
to July 3. Ref. No. ALINORM 63/12 July 1963. Rome:
Food and Agriculture Organization of the United
Nations.

FAO/WHO (1964) Evaluation of Toxicity of Pesticide
Residues of Food. Report of the Joint Meeting of the
FAO Committee on Pesticides in Agriculture and the WHO
Expert Committee on Pesticide Residues. FAO Meeting
Report No. PL/1963/13. WHO/Food Add./23.

FAO/WHO (1967a) Evaluation of Some Pesticide Residues in
Food. Pages 188-203, Joint Meeting of the FAO Working
Party and the WHO Expert Committee on Pesticide
Residues, Geneva, November 14-21, 1966. FAO, PL:CP/15,
WHO/Food Add./67. 32. Geneva: World Health
Organization.

FAO/WHO (1967b) Pesticide Residues in Food. Joint Report
of the FAO Working Party of Pesticide Residues and the
WHO Expert Committee on Pesticide Residues, Geneva,
November 14-24, 1966, FAO Agricultural Studies No.,
73, WHO Tech. Rept. Ser. No. 370. Geneva: World Health
Organization.

FAO/WHO (1967c) Specifications for the Identity and
Purity of Food Additives and Their Toxicological
Evaluation: Some Emulsifiers and Stabilizers and
Certain Other Substances. Tenth report of the Joint
FAO/WHO Expert Committee on Food Additives, Geneva,
October 11-18, 1966. FAO Nutrition Meetings Rept. Ser.
No. 43, WHO Tech. Rept. Ser. No. 373. Geneva: World
Health Organization.

FAO/WHO (1968a) 1967 Evaluations of Some Pesticide
Residues in Food. Pages 200-209, Joint Meeting of the
FAO Working Party of Experts and the WHO Expert
Committee on Pesticide Residues. Rome, December 4-11,
1967. FAO/PL:1967/M/11/1, WHO/Food Add./68.30. Geneva:
World Health Organization.

FAO/WHO (1968b) Pesticide Residues. Report of the 1967
Joint Meeting of the FAO Working Party and the WHO
Expert Committee. Rome, December 4-11, 1967. FAO
Meeting Report No. PL:1967/M/11, WHO Tech. Rept. Ser.
No. 391. Geneva: World Health Organization.

FAO/WHO (1969) Pesticide Residues in Food. Joint Report
of FAO Working Party of Experts on Pesticide Residues
and the WHO Expert Committee on Pesticide Residues.
Geneva, December 9-16, 1968, FAO Agricultural Studies
No. 78, WHO Tech. Rept. Ser. No. 417. Geneva: World
Health Organization.

FAO/WHO (1970) Pesticide Residues in Food. Joint Report
of FAO Working Party of Experts on Pesticide Residues
and the WHO Expert Group on Pesticide Residues. Rome,

December 8-15, 1969, FAO Agricultural Studies No. 84, WHO Tech. Rept. Ser. No. 458. Geneva: World Health Organization.

FAO/WHO (1971) Pesticide Residues in Food. Joint Report of FAO Working Party of Experts on Pesticide Residues and the WHO Expert Group on Pesticide Residues. Rome, November 9-16, 1970, FAO Agricultural Studies No. 87, WHO Tech. Rept. Ser. No. 474. Geneva: World Health Organization.

FAO/WHO (1972a) Pesticide Residues in Food. Joint Report of FAO Working Party of Experts on Pesticide Residues and the WHO Expert Group on Pesticide Residues. Geneva, November 22-29, 1971, FAO Agricultural Studies No. 88, WHO Tech. Rept. Ser. No. 502. Geneva: World Health Organization.

FAO/WHO (1972b) Evaluation of Mercury, Lead, Cadmium, and the Food Additives Amaranth, Diethylpyrocarbonate, and Octyl Gallate. Joint FAO/WHO Expert Committee on Food Additives, WHO Food Additives Series No. 4. Geneva: World Health Organization.

FAO/WHO (1972c) Evaluation of Certain Food Additives and the Contaminants Mercury, Lead, and Cadmium. Sixteenth report of the Joint FAO/WHO Expert Committee on Food Additives. Geneva, April 4-12, 1972, FAO Nutrition Meetings Rept. Ser. No. 51, WHO Tech. Rept. Ser. No. 505. Geneva: World Health Organization.

FAO/WHO (1973) Pesticide Residues in Food. Joint Report of FAO Working Party of Experts on Pesticide Residues and the WHO Expert Group on Pesticide Residues. Rome, November 20-28, 1972, FAO Agricultural Studies No. 90, WHO Tech. Rept. Ser. No. 525. Geneva: World Health Organization.

FAO/WHO (1974a) List of Maximum Levels Recommended for Contaminants by the Joint FAO/WHO Codex Alimentarius Commission. First Series, Joint FAO/WHO Food Standards Programme, Codex Alimentarius Commission, CAC/FAL 2-1973. Rome: Food and Agriculture Organization of the United Nations.

FAO/WHO (1974b) The Use of Mercury and Alternative Compounds as Seed Dressings. Joint Report of FAO/WHO, Geneva, March 4-6, 1974, FAO Agricultural Studies No. 95, WHO Tech. Rept. Ser. No. 555. Geneva: World Health Organization.

FAO/WHO (1974c) Pesticide Residues in Food. Joint Report of the FAO Working Party of Experts on Pesticide Residues and the WHO Expert Committee on Pesticide Residues. Geneva, November 26-December 5, 1973, FAO Agricultural Studies No. 92, WHO Tech. Rept. Ser. No. 545. Geneva: World Health Organization.

FAO/WHO (1975) Pesticide Residues in Food. Joint Report of the FAO Working Party of Experts on Pesticide Residues and the WHO Expert Committee on Pesticide

Residues. Rome, December 2-11, 1974, FAO Agricultural
Studies No. 97, WHO Tech. Rept. Ser. No. 574. Geneva:
World Health Organization.
FAO/WHO (1976) Pesticide Residues in Food. Joint Report
of the FAO Working Party of Experts on Pesticide
Residues and the WHO Expert Committee on Pesticide
Residues. Geneva, November 19-22;, 1975. FAO Plant
Production and Protection Series No. 1, WHO Tech.
Rept. Ser. No. 592. Geneva: World Health Organization.
Goldwater, L.J. (1974) Standards and Regulations for the
Control of Mercury in the Environment. Pages 205-212,
Proceedings of the 1st Congress Internacional del
Mercurio, Barcelona, 6-10 de Mayo de 1974. Tomo II.
Fabrica Nacional de Moneda y Timbre. Madrid, Spain.
WHO (1976a) Conference on intoxication due to
alkylmercury-treated seed. Bull. WHO Suppl. 53:1-138.
WHO (1976b) Environmental Health Criteria 1, Mercury.
Report from a meeting held 4-10 February, 1975.
Geneva: World Health Organization.

APPENDIX C

CURRENT STATUS OF STATE SPORT AND COMMERCIAL
FISHERIES WITH RESPECT TO MERCURY POLLUTION
JUNE 1977

The following states have reported no closures of
sport and/or commercial fisheries and have not issued
any restrictions or health warnings regarding the
consequences of eating mercury-contaminated fish and
other seafood:

Alaska	Indiana	New Jersey
Arizona	Iowa	North Dakota
Arkansas	Kansas	Oklahoma
Colorado	Maine	Puerto Rico
Connecticut	Maryland	Rhode Island
Delaware	Missouri	Utah
Florida	Montana	Washington
Guam	Nebraska	Wyoming
Hawaii	Nevada	

The listing below describes the current status of the
states that have, since 1970, closed sport or commercial
fisheries and/or issued health warnings about the
consequences of eating fish or other seafood
contaminated with mercury. The status of mercury in the
United States as a whole is illustrated in the
attachment at the end of this Appendix.

STATE	CURRENT STATUS
Alabama*	The 1970 restriction on commercial fishing in the Tombigbee, Tensaw, and Mobile rivers and their respective tributaries as well as the waters of upper Mobile Bay was lifted on July 7, 1972. However, all of the Pickwick Reservoir in Alabama still remained closed. The Pickwick impoundment closed to commercial fisheries on July 7, 1970,

STATE	CURRENT STATUS

was reopened on May 21, 1975. This closure was not mentioned in the 1970 FWQA Survey.[1] Alabama has not issued a health warning with respect to the dangers of eating mercury-contaminated fish.

California**   The warning to eat only one meal per week of striped bass and catfish from the Sacramento-San Joaquin Delta and San Francisco Bay area were issued by the State Department of Health and are still in effect. In addition, in 1972, warnings were issued by the Santa Clara County Park and Recreation District that fish (largemouth bass, sunfish, catfish, and rainbow trout) taken from Calero, Almaden, and Guadalupe reservoirs may contain high levels of mercury and should not be eaten.

Georgia*   In 1970 the Savannah River and New Savannah Dam on Highway 17 as well as the Brunswick Estuary were closed to sport fishing. The Brunswick Estuary was also closed to commercial fishing. All restrictions and closures were removed from the Brunswick Estuary on October 19, 1970, and the Savannah River in September, 1972.

Idaho**   No state restrictions or fishery closure. Conditional warnings (no person should eat more than 1/2 lb of fish per week; and pregnant women, infants, and children should not eat any fish taken from American Falls Reservoir) were issued by the State Health Department for selected species of fish in the American Falls Reservoir (January, 1971 and 1972), Hells Canyon Dam, and other reservoirs on the Snake River (January 1971).

Illinois**   The 1970 policy of no state sport or commercial fishery closures and no health warning advice to fishermen or the public about the consequences of eating mercury-contaminated fish was revised because certain species of fish taken from three reservoir lakes (Rend Lake, Cedar Lake, and Lake Shelbyville) exceeded the FDA mercury action

[1]Harlan, J.R. (1971) Mercury Pollution Survey. Sport Fishing Institute Bulletin No. 221. Washington, D.C.: 719 13 Street, N.W.

STATE	CURRENT STATUS

level of 0.5 μg/g. As a result, the public was warned to limit consumption to no more than 1/2 lb per week of largemouth bass, shorthead redhorse, black buffalo, bullhead, and yellow bullhead from these lakes. No mercury contamination problems have been identified that affect the commercial fisheries.

Kentucky**   The 1970 health warning and restrictions issued for fish taken from the Tennessee River at Calvert, Kentucky, have been relaxed. At present, a warning pertains only to local residents who eat 20 meals per week of fish from the watercourse over a long period of time (years). Nonresidents are not affected by this warning.

Louisiana*   In 1970 Louisiana issued a health warning for fish taken from the Calcasieu River and stopped the interstate shipment of these fish. All state restrictions were removed in 1975 because the mercury concentrations in fish were within the FDA tolerance level of 0.5 μg/g.

Massachusetts**   In 1970, minor fishery closures and a health warning were issued for three specific areas and fisheries. As a result of mercury contaminations above the FDA action level of 0.5 μg/g, two shellfish areas were closed in December 1970. The areas were Sippican Harbor in Marion and Quisset Harbor in Falmouth. In 1975, the total closures were modified and portions of the harbors were reopened to shellfishing because mercury levels had declined. Neither area was heavily industrialized, and the source of mercury was marinas that had utilized mercury-based paint in boat yard work. Also in 1970, a health warning was issued for persons who were engaged in recreational finfishing in the Taunton River. Fish could be taken, but people were advised not to consume those from the northern boundary of the town of Fall River north to the northern boundary of the town of Dighton. This warning was a result of an industrial discharge, and the advisory is still in effect. The industrial discharge has since been terminated.

STATE	CURRENT STATUS
Michigan***	On April 15, 1970, sport fishing was banned and health warnings posted on the St. Clair River and Lake St. Clair. The commercial fishing for walleye in Lake Erie was banned on April 29, 1970. On May 20, 1970, the sport fishing restrictions were reduced to "catch and release" status in the St. Clair and Detroit rivers and Lake St. Clair. In Lake Erie and Lake Huron (south of Port Sanilac) sport fishermen could keep all fish except walleye, white bass, and freshwater drum while commercial fishermen could keep all species except walleye. During the summer of 1970 the courts set aside the sport fishing ban and allowed fishermen to keep their catches. This ruling is still in effect as is the health warning originally imposed in the spring of 1970. Subsequently, walleye were given sport status, which protected them from commercial exploitation.
Minnesota**	There have been no closures of sport or commercial fisheries in the state. On December 11, 1970, the Department of Health advised that anglers restrict intake of fish from certain water to once a week due to high mercury levels. Subsequent analyses of fish for mercury resulted in several modifications of the warning between 1970 and 1976. The following four watercourses were found to contain some fish exceeding the FDA action level of 0.5 $\mu g/g$: (1) the St. Louis River below Coloquet, (2) the Upper Mississippi River between Grand Rapids and Brainerd, (3) the Red River along the Dakota border, and (4) Crane Lake near the Canadian border. The latest modification was on May 14, 1976, when it was advised that fish from Crane Lake be eaten no more than once a week.
Mississippi**	On August 1, 1975, the Mississippi portions of Pickwick Lake were reopened to commercial fishing. In addition, all fish containing more than 0.5 $\mu g/g$ mercury would be subjected to seizure by the Food and Drug Administration. A warning was also issued by the Mississippi State Board of Health that eating fish from Pickwick Lake by pregnant women should be kept to a minimum and that all other persons should restrict their

STATE	CURRENT STATUS

normal intake of fish from this lake to not more than two meals per week.

New Hampshire* No state restrictions. The 1970 danger warnings for pickerel, yellow perch, and smallmouth bass from the Merrimack and Connecticut rivers have been removed because public health officials contend that the current creel limits preclude anyone from eating sufficient quantities of fish to be harmful to health.

New Mexico** Sport fishery health cautions for the Navajo and Ute Lakes were issued by the Health and Social Services Department (HSSD) in 1970 and are still in effect. The public was advised not to eat more than 2 lb per week of any species of fish taken from Navajo Lake. If walleye and largemouth bass larger than 1.5 lb were taken from Ute Lake, the recommended consumption was to be limited to less than 1 lb per week per adult person, and the recommended consumption of catfish larger than 5 lb was limited to 2 lb per week per person. Warnings against eating large amounts of fish taken from Summer, Elephant Butte, and Caballo lakes were also issued. The HSSD stressed that it was safe to eat fish from any New Mexico lake provided that the recommended consumption limits were observed. No action was taken by the State Game and Fish Department pursuant to the HSSD warnings. Very little public concern has been evident since 1971.

New York** With the exception of three bodies of water, officials have proclaimed that it is safe to eat fish once a week without fear of mercury contamination. Onondaga Lake is closed to fishing. People are advised not to eat lake trout from Lake George or muskellunge from the St. Lawrence River, although all other species of fish from these waters may be eaten once a week. Pregnant women and infants are advised not to eat any freshwater fish. There are no commercial fishing restrictions.

North Carolina** For the inland fisheries, the 1970 general danger warnings to fishermen are still in effect. No closures or health warnings have been issued for the marine fisheries. However, the FDA ban on swordfish

STATE	CURRENT STATUS

	ended a small fishery for this species on the northern coast of this state.
Ohio*	In 1970 the Lake Erie commercial fishery was closed for all fish except perch. An embargo was placed on white bass and a sport fishery health warning announced. Since then the 1970 restrictions were ruled unconstitutional by the Ohio Supreme Court because "The Division of Wildlife is not and was not responsible for consumer protection." No state restrictions or health warnings are presently in effect.
Oregon*	The 1970 policy of no state or commercial fishery closures continues. However, in 1970 health warnings were issued for rainbow trout, black crappie, suckers, and largemouth bass taken from the Antelope and Owyhee reservoirs and parts of the Wallametta River. A curtailed intake of any fish taken from these waters was recommended particularly for infants and pregnant women. In 1975 a similar warning was issued for striped bass.
Pennsylvania*	In 1970 the Department of Environmental Resources issued an advisory that large predator game fish such as walleye, drum, smallmouth bass, and white bass may exceed the FDA action level of 0.5 μg/g for mercury, and therefore some restriction on the human consumption of these fishes may be advisable. At present, Pennsylvania officials do not consider mercury pollution a serious problem in their state and have not placed any official restrictions on catching game fish. Nor have any health warnings been issued with respect to eating the species.
South Carolina**	In 1970 the sport and commercial fisheries were closed on the Savannah River from Augusta, Georgia, to the coast. These restrictions were removed in 1972. In 1972 an advisory was issued that recommended limiting the consumption of fish taken from Lake Jocassee to 1.5 lb of dressed fish per week and no intake by pregnant women. The elevated levels of mercury in Lake Jocassee fish were the result of natural conditions. They included the slightly higher soil mercury levels in the lake area and, more

STATE	CURRENT STATUS
	significantly, the oligotrophic condition of the lake. The advisory is currently in effect, and the mercury levels are being monitored.
South Dakota**	In 1970 the state reported no closures or advice to fishermen about health hazards associated with eating fish taken from South Dakota waters. Since then, only the Cheyenne Arm of Oahe Reservoir has been posted by the state's health officer. In June 1973, commercial and sport fishermen were warned not to eat more than 1 1/2 lb of fish from this water per week. This health warning is still in effect.
Tennessee***	In September 1970, the the Tennessee River and Pickwick Lake commercial fisheries were closed and a health warning along with a catch and release policy instituted for their sport fisheries. Both the commercial and sport fisheries restrictions were removed from Pickwick Lake and the Tennessee River in August 1971. The catch and release restrictions and health warning imposed on sport fishing in the North Fork Holston River in September 1970 are still in effect and commercial fishing is not allowed.
Texas***	In 1970 approximately 19,900 acres of Lavaca Bay was closed to commercial oyster harvest. As of September 1, 1971, the number of acres of Lavaca Bay that are closed to the commercial harvest of shellfish has been reduced from 19,900 to 11,000 acres. Additionally, the acreage closed to the harvest of oysters was not closed in its entirety because of mercury pollution. Prior to its reclassification in 1970, Lavaca Bay had approximately 8500 acres which were closed because of sanitary and bacteriological reasons.
Vermont***	In 1970 Lake Champlain and its tributaries were closed to the commercial harvest of walleye. In addition, an embargo on commercial sales of walleye from Lake Champlain, its tributaries, and Lake Memphremagog is still in effect. On April 25, 1973, the sport fishery danger warnings imposed in 1970 were continued for the consumption of walleye from Lakes Champlain, its tributaries, and Lake Memphremagog.

STATE	CURRENT STATUS
Virginia***	In 1970 the sport fishery on the North Fork of the Holston River below Saltville was closed by the Virginia Department of Health. In 1975 this restriction was relaxed to permit fishing under a catch and release regulation. The health warning issued in 1970 and again in 1975 concerning the danger of eating fish taken from these waters is still in effect. On June 6, 1977, the Virginia Department of Health closed the sport fishery on the South River, the south fork of the Shenandoah River. Sport fishing is allowed under a "catch and release" policy, but citizens are warned that fish from these waters are unfit for human consumption.
West Virginia*	Sport and commercial fisheries in West Virginia are presently not restricted due to mercury pollution. The Ohio River commercial fishery which was closed on August 20, 1970, was reopened on July 1, 1973. Currently, West Virginia has no health warnings about the dangers of eating mercury-contaminated fish.
Wisconsin*	In 1970 a catch and release policy was recommended for the Wisconsin River along with a health warning not to consume more than one meal per week of fish taken from this river. At present there are no state restrictions because mercury levels in the Wisconsin River system have recently dropped below the FDA action level of 0.5 $\mu g/g$. Contracts for commercial fishing are now being granted for the Wisconsin River and its impoundments, and warnings on fish consumption limits are no longer being issued.

 * States that have rescinded closures of sport and/or commercial fisheries or health warnings issued since 1970.

 ** States where health warnings are in effect about the consequences of eating mercury-contaminated fish or other seafood from selected watercourses in the state.

*** States where sport or commercial fisheries are currently closed and health warnings are in effect.

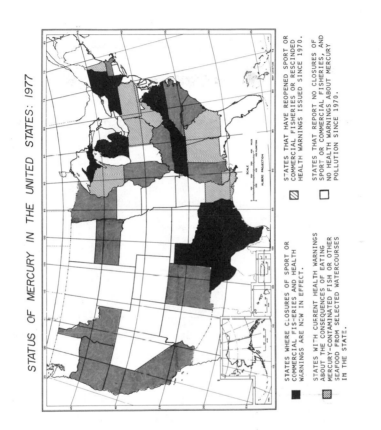

STATUS OF MERCURY IN THE UNITED STATES: 1977

STATES WHERE CLOSURES OF SPORT OR
COMMERCIAL FISHERIES AND HEALTH
WARNINGS ARE NOW IN EFFECT.

STATES WITH CURRENT HEALTH WARNINGS
ABOUT THE CONSEQUENCES OF EATING
MERCURY-CONTAMINATED FISH OR OTHER
SEAFOOD FROM SELECTED WATERCOURSES
IN THE STATE.

STATES THAT HAVE REOPENED SPORT OR
COMMERCIAL FISHERIES OR RESCINDED
HEALTH WARNINGS ISSUED SINCE 1970.

STATES THAT REPORT NO CLOSURES OF
SPORT OR COMMERCIAL FISHERIES, AND
NO HEALTH WARNINGS ABOUT MERCURY
POLLUTION SINCE 1970.